Astronomers' Universe

More information about this series at
http://www.springer.com/series/6960

Rhodri Evans

The Cosmic Microwave Background

How It Changed Our Understanding of the Universe

 Springer

Rhodri Evans
School of Physics & Astronomy
Cardiff University
Cardiff
United Kingdom

ISSN 1614-659X ISSN 2197-6651 (electronic)
ISBN 978-3-319-09927-9 ISBN 978-3-319-09928-6 (eBook)
DOI 10.1007/978-3-319-09928-6
Springer Cham Heidelberg New York Dordrecht London

Library of Congress Control Number: : 2014957530

© Springer International Publishing Switzerland 2015
This work is subject to copyright. All rights are reserved by the Publisher, whether the whole or part of the material is concerned, specifically the rights of translation, reprinting, reuse of illustrations, recitation, broadcasting, reproduction on microfilms or in any other physical way, and transmission or information storage and retrieval, electronic adaptation, computer software, or by similar or dissimilar methodology now known or hereafter developed. Exempted from this legal reservation are brief excerpts in connection with reviews or scholarly analysis or material supplied specifically for the purpose of being entered and executed on a computer system, for exclusive use by the purchaser of the work. Duplication of this publication or parts thereof is permitted only under the provisions of the Copyright Law of the Publisher's location, in its current version, and permission for use must always be obtained from Springer. Permissions for use may be obtained through RightsLink at the Copyright Clearance Center. Violations are liable to prosecution under the respective Copyright Law.
The use of general descriptive names, registered names, trademarks, service marks, etc. in this publication does not imply, even in the absence of a specific statement, that such names are exempt from the relevant protective laws and regulations and therefore free for general use.
While the advice and information in this book are believed to be true and accurate at the date of publication, neither the authors nor the editors nor the publisher can accept any legal responsibility for any errors or omissions that may be made. The publisher makes no warranty, express or implied, with respect to the material contained herein.

Cover illustration: The Orion Nebula, the closest of over 100 known huge gas and dust clouds located in our Milky Way Galaxy, appears to be a very busy "newborn nursery" where many new solar systems are being built. This photograph courtesy of NASA/ESA/Luca Ricci shows a collection of six such new planetary systems, some of which may in the far distant future, provide a home for life. Credit: NASA, ESA, M. Robberto (Space Telescope Science Institute/ESA), the Hubble Space Telescope Orion Treasury Project Team and L. Ricci (ESO)

Printed on acid-free paper

Springer is part of Springer Science+Business Media (www.springer.com)

For my parents Colin and Audrie; my sister Jill, my wife Maggie and my children Meirin, Siân-Azilis and Esyllt. Thank you.

Preface

On Sunday the 16th of March this year, just a few weeks ago from my writing this, rumours started circulating on social media that there would be a *major* announcement from cosmologists at Harvard University the following day. By the following morning it was clear that the announcement was going to be coming from the team conducting an experiment at the South Pole using a telescope called BICEP2. When the press conference arrived the physics and cosmology communities almost went into a meltdown. The BICEP2 team announced that they had found the best evidence yet for something called "cosmic inflation", which happened when the Universe was 0.0000000000000000000000000000000001 of a second old.

Although this result is currently hotly disputed, if correct, it is a spectacular evidence that we understand the physics of the Universe as far back as this unimaginably small fraction of time after it came into existence. However, since the March 2014 announcement by the BICEP2 team, a huge amount of debate has ensued in the cosmology community as to whether their result is real or not. The arguments in this debate will be discussed in more detail in the book.

"*Cosmology*" is the subject of understanding the beginning, evolution and nature of the Universe. It is a subject which has always fascinated humanity, and every civilisation has its own cosmology. Whether it is the "*Genesis story*" common to Judaism, Christianity and Islam; the Hindu's belief that Lord Brahma created the Universe, Lord Vishnu maintains it, and Lord Shiva will destroy it; or the complex "*dream time*" origin stories of the Australian Aboriginals, as a species we have always tried to understand the Universe, and from where it and we ourselves have come. It seems to be a basic human need in all of us. It is also a subject which has fascinated me ever since I was 12 and saw a BBC Horizon programme called "*The Key to the Universe*".

In this book I present scientists' current favourite cosmology, the "*hot big bang theory*". This theory postulates that our Universe began some 13.7 billion years ago when time and space were created in an unimaginably dense "fireball"—the big bang. Since that moment of creation, space has been expanding and evolving. The stars, galaxies, elements and life in our Universe have all developed from this.

This book traces the history of our understanding of the Universe, from the early ideas of the Greeks through to the latest findings announced in the last few weeks which probe the conditions in the very earliest moments of our Universe's existence. After laying down the evidence that our Earth is not the centre of the Universe as the Greeks had thought, but rather orbiting the Sun on the outskirts of an average galaxy, I show how we now know that our Milky Way is just one of billions of galaxies, and that the Universe is expanding. I present the story of the 1931 prediction of the big bang, and the 1948 prediction of a relic radiation from the early Universe which we call the "*cosmic microwave background radiation*".

This radiation was finally discovered in 1964, and since then advances in both theory and observations (such as the BICEP2 experiment mentioned above) now allow us to argue that we understand the physics of the Universe back to the briefest fraction of a second after it started (we are still not quite sure about the time before that!).

What a remarkable journey we have made; barely 100 years ago we did not know how stars got their energy, whether our Milky Way was the entire Universe, or from where the elements which make up our very fabric had come. We now believe we know all of these things. Yet there is still so much that we don't understand. Understanding the details of the structure of the Universe has led us to the remarkable and exciting finding that ordinary matter may only make up about 5 % of the Universe; with the rest being made up of about 20 % of "dark matter" and about 75 % of "dark energy". At this present time, we have no idea what dark matter and dark energy are, just that it seems they exist.

My aim is that this book should be of interest to all those people who want to learn more about where our Universe has come from, and how we have gone from the ignorance of the past to the understanding we have today. I have tried to make the story as engaging as possible, science is a human activity and I have attempted to give a flavour of some of the people who have played key parts in advancing our knowledge of the Universe. I have also tried to explain any necessary physics at a level which I hope will be understandable to a non-scientist; any lack of understanding by you the reader is due to my own inadequacies in doing a proper job in my explanations.

We stand at a remarkable moment in *cosmology*. In the last 30 years separate lines of evidence have come together to give us a consistent picture of the origin, composition and structure of the Universe. However, there are also many unanswered mysteries including dark matter, dark energy and whether our Universe is just part of a "*multiverse*". There is, however, a caveat; as my ex-Ph.D. supervisor and "science mentor" Professor Mike Disney always reminds me, there has never been a time in history when people didn't think that they had the correct cosmology. Maybe our current cosmology is as incorrect as our ancestors'. Only time will tell.

Wales, UK   Rhodri Evans
April 2014

Contents

1. **At the Centre of Creation?** 1
   1.1 Ptolemy's Universe 1
   1.2 Copernicus' Revolution 3
   1.3 The Man with the Metal Nose 5
   1.4 Dancing Moons and Crescent Planets 11
   1.5 On the Shoulders of Giants 15
   1.6 The Relative Size of the Solar System 17
   1.7 Edmund Halley, le Gentil, Captain Cook and Sex for Iron Nails 18
   1.8 Do the Stars Move? 26
   1.9 Mapping the Milky Way and Nebulae 26
   1.10 Stellar Fingerprints 28
   1.11 Cepheid Variables and Standard Candles 29
   References 31

2. **A Universe of Galaxies** 33
   2.1 Kapteyn's Universe 33
   2.2 Galaxies Rushing Away 37
   2.3 Herber Curtis, Harlow Shapley, and the Great Debate 41
   2.4 Blinking Stars in the Andromeda Nebula 45
   2.5 Hubble, Humason and an Expanding Universe 50
   References 53

3. **The Cosmic Microwave Background** 55
   3.1 The Primordial Atom 55
   3.2 Fred Hoyle and the Steady State Theory 59
   3.3 Modern Alchemy 60
   3.4 The Origin of the Chemical Elements 61
   3.5 The Hot Surfaces of Stars 63
   3.6 Cooking Helium in the Early Universe 65
   3.7 Cosmological Model Predictions 66
   3.8 The First Prediction of the Cosmic Microwave Background 69
   3.9 From Where Does This Microwave Background Come? 71

	3.10	A New Window on the Heavens	72
	3.11	Missed Opportunities	76
	3.12	Clearing Out the Pigeon Droppings	78
	3.13	Was It Really a Blackbody?	84
	References		86
4	**The Cosmic Background Explorer (COBE)**		91
	4.1	The Earth, Sun and Milky Way's Motions in Space	91
	4.2	Invisible Matter	95
	4.3	Filaments and Voids	99
	4.4	Cosmic Inflation	100
	4.5	Studying the CMB from Space	105
	4.6	Waiting for COBE	108
	4.7	A Perfect Blackbody	110
	4.8	The "Scientific Discovery of the Century"	112
	References		115
5	**To the Ends of the Earth**		117
	5.1	Primordial Sound Waves	117
	5.2	Looking for the First Acoustic Peak	121
	5.3	The Most Southerly Continent	122
	5.4	"Great God! This Is an Awful Place"	127
	5.5	The Center for Astrophysical Research in Antarctica	130
	5.6	The Universe is Flat!	135
	5.7	Polarisation in the Cosmic Microwave Background	136
	References		141
6	**The Evidence Comes Together**		143
	6.1	Back into Space	143
	6.2	The Cosmic Web	155
	6.3	A New Standard Candle	157
	6.4	The Most Surprising Astronomical Finding of the Century?	159
	6.5	MAP's Detailed Image of the Baby Universe	160
	References		165
7	**Up to the Present, and Beyond**		169
	7.1	Europe Looks Back to the Dawn of Time	169
	7.2	Searching for Gravitational Waves	178
	7.3	Neutrino Astrophysics	181
	7.4	Can We Learn Much More from Studying the CMB?	185
	7.5	What Is Dark Matter?	194
	7.6	What Is Dark Energy?	195
	7.7	The Moment of Creation	195
	7.8	The Cosmic Onion	197
	Concluding Remarks		198
	References		198
Glossary			201

Chapter 1
At the Centre of Creation?

Early models of the Universe by Greeks such as Ptolemy placed the Earth at the centre of the heavens. In the sixteenth century, Copernicus suggested that the Sun and not the Earth was at the centre of creation, and Galileo found observational evidence for this in the 1609–1611 period. Kepler worked out that the planets orbit the Sun in ellipses and not circles, and later in the seventeenth century, Newton wrote down the laws of gravity and mechanics which shaped our understanding of Physics for the next 250 years. Using a method suggested by Halley, in 1761 and 1769 we measured the distance from the Earth to the Sun, and by 1838 we had measured the distance to one of the nearest stars. In 1814 Fraunhofer learnt that stars have dark lines in their spectra, and this could be used to measure how quickly they are moving towards or away from us, and by the 1910s Henrietta Leavitt had discovered a way to measure vast distances using a particular type of star called a Cepheid variable.

1.1 Ptolemy's Universe

To even the casual observer, the daily motion of the Sun from East to West across the sky is very obvious. A little more observations over a few weeks and one will notice the changing phases of the Moon, with each full Moon spaced by about 30 days. Beyond this, more detailed observations of the night-time sky can reveal that different stars appear at different times of the year. For example, the easily recognisable stars that form the constellation Orion can be seen in the autumn and winter skies, but not in the summer. Conversely, the stars Vega, Deneb and Altair which form the summer triangle cannot be seen in the winter skies.

Even more detailed observations will show that not all the stars behave the same way. There are five star-like objects which appear to wander amongst their fellow stars. These are the five naked eye planets, Mercury, Venus, Mars, Jupiter and

Saturn. The Greek name 'planet' actually means 'wandering star'. In any model of the Universe, these observations need to be explained.

The model which has been believed for most of the last 2,000 years is one that is chiefly attributed to the Greek astronomer Claudius Ptolemy [1], who lived in the second century BC in Alexandria. In his model, the Earth was placed at the centre, with the Moon, Sun, planets and stars all orbiting it at different rates.

A model is, of course, only successful if it can match the observations. Although the planets appear to normally wander through the background stars in an easterly direction from night to night and month to month, sometimes they appear to reverse direction. The planet which shows this brief westerly motion against the background stars most markedly is Mars. This phenomenon, called 'retrograde motion', could not be explained with the simple model shown above. Ptolemy had to introduce the idea of 'epicycles', which were superimposed on the planet's motion along its deferent (the circles shown in Fig. 1.1). By adjusting the size of a planet's epicycle,

Fig. 1.1 Ptolemy's model of the Universe placed the Earth at the centre, with the Sun, Moon, planets and stars all moving about it. This drawing is taken from Peter Aplan *"Cosmographia"* (1524)

Ptolemy was able to get excellent agreement between the positions of celestial objects as predicted by his model and what was observed.

Ptolemy's incorrect 'geocentric' (Earth-centred) model was not overthrown until the early 1600s. The Polish astronomer Nicholas Copernicus is most often credited with suggesting that the Sun and not the Earth lay at the centre of the Universe. However, he was not the first. Even before Ptolemy, another Greek astronomer, Aristarchus of Samos, working in the third century BC proposed that the Sun and not the Earth was at the centre of the Universe. But Ptolemy's model was preferred, mainly because it fitted in well with the philosophical world-view developed by Plato and Aristotle.

1.2 Copernicus' Revolution

The person who is most credited with suggesting the heliocentric (Sun-centred) Universe is Nicholas Copernicus. Copernicus had trained as a church cleric, and through the help of his uncle who was Bishop of Ermland, he held a position as a canon at the cathedral in Frauenburg in modern-day Poland. Copernicus had studied law and medicine in Italy, so his main duties as a canon were to act as physician and secretary to his uncle. His duties were not particularly demanding, allowing him to indulge in his real passion, astronomy.

In 1514, Copernicus wrote a 20 page pamphlet *Commentariolus* [2] ('Little Commentary') which laid out his belief about the Earth's place in the Universe. Although it was not formally published, he circulated it amongst a few people, and in it he laid out some of the most radical ideas that anyone had made. He stated that the heavenly bodies do not contain a common centre and that the Earth does not lie at the centre of the Universe. Rather, he said, the Sun lay near the centre of the Universe, and the distance from the Earth to the Sun was tiny compared to the distances to the stars. He also stated that the apparent motion of the Sun and stars across the sky from East to West was due to the Earth's rotation, and that the change in the position of the Sun from season to season was due to our motion around it. Finally, he stated that the apparent retrograde motion of some of the planets was due to seeing their motion from a moving Earth.

He was correct in each one of these statements, something to which very few astronomers of his age would have agreed, but which most would do by the end of the sixteenth century. Around the time that Copernicus wrote this pamphlet, his uncle died, and Copernicus moved to Frauenberg Castle where he set up his own observatory. He spent the rest of his life reworking Commentariolus into a much more learned and complete piece of work. With his religious training as a canon, Copernicus was fully aware how his arguments flew in the face of hundreds of years of Church teaching, and so was reluctant to share his work with anyone for fear of persecution.

However, in 1539 a young German man by the name of Georg Joachim von Lauchen, but known as Rheticus, travelled from Wittenberg to meet Copernicus.

Rheticus spent 3 years at Frauenberg Castle with Copernicus, reading his manuscript and giving him feedback and reassurance on its contents. By 1541, Rheticus had received permission from Copernicus to take the manuscript to a printing house in Nuremberg for publication. At last, in the spring of 1543, *De revolutionibus orbium caelestium* [3] ('On the Revolutions of Heavenly Spheres') was published. But, in late 1542 Copernicus had suffered a cerebral haemorrhage, and had spent the intervening months in bed struggling to stay alive. Copies of his book reached him just as he was losing his battle, and upon seeing his life's work finally in print his life slipped away. His friend Canon Giese wrote a letter to Rheticus saying

> For many days he had been deprived of his memory and mental vigour; he only saw his completed book at the last moment, on the day he died.

Copernicus' model placed the Sun at the centre of the Universe, with the Earth, Moon, planets and stars orbiting the Sun in perfectly circular orbits (see Fig. 1.2). Mercury and Venus were placed inside of Earth's orbit, the Moon orbiting the Earth as Earth orbited the Sun, and Mars, Jupiter and Saturn orbited outside of Earth's orbit. Finally, the last sphere in this cosmic Russian doll was the sphere of the 'fixed stars'.

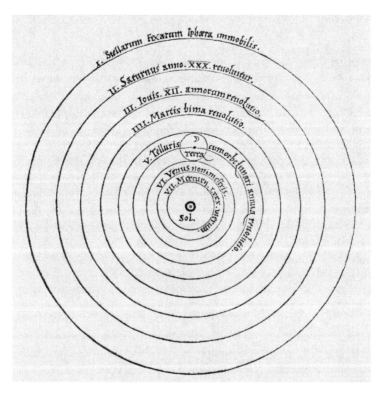

Fig. 1.2 Copernicus' model of the Universe placed the Sun and not the Earth at the centre (image from Copernicus' *De revolutionibus orbium coelestium* (1543))

Much like Aristarchus before him, Copernicus' Heliocentric model was quickly dismissed as it was found to be less accurate than Ptolemy's. In some ways it was not surprising that Ptolemy's model was so successful; over the centuries the size and speed of the planets' motions along their deferents, and the size of the epicycles, had been refined to agree with observations. It had become a very successful model in accurately predicting the positions of the planets, but it had also become highly complex in order to achieve this accuracy. Copernicus' model could not have been more different, it was elegant in its simplicity. It simply had the Sun at the geometric centre of a series of perfect spheres, with the other celestial bodies and the Earth moving within these spheres.

1.3 The Man with the Metal Nose

The most able observational astronomer of the late 1500s was also its most colourful, Tycho Brahe. Born into Danish nobility in 1546, at the tender age of twenty he got involved in a duel with his cousin Manderup Parsberg and had the bridge of his nose sliced off. From this point on, Brahe glued a false metal nose in its place, but by carefully blending gold and silver and copper the colour actually matched his skin tone and most people were unaware of the prosthetic.

But, although eccentric, Brahe is best known for the accuracy of his astronomical observations. His reputation led to King Frederick II of Denmark giving him his own island, Hven, 10 km off the Danish coast near Copenhagen, and paid for Brahe to build an observatory there. Brahe named his observatory *Uraniborg*, which means 'Castle of the Heavens', and one could argue that it was the most expensive observatory ever built as it used up more than 5 % of Denmark's gross national product!

The Observatory was lavishly furnished with a printing press, a library, accommodation for the servants who would help Brahe with his observations, and observing towers equipped with the best instruments of the day. Remember, this is before the invention of the telescope; but the sextants, quadrants and other naked-eye observing tools were of a better accuracy and quality than any that had been used before.

With these facilities at his disposal, Brahe produced the most accurate stellar and planetary positions ever seen. He would typically measure the position of celestial objects to an accuracy of 1/30th of a degree (for comparison, the full Moon is 1/2 of a degree across). This was about five times better than anyone had previously obtained. As Brahe's fame spread, a steady stream of important people came to Uraniborg to visit him. In addition to wishing to see the impressive observatory and its eccentric director, they were probably also drawn by the wild parties that Brahe threw there (Fig 1.3).

His observations found disagreement with the Ptolemaic model. Brahe had a copy of Copernicus' De revolutionibus in his library and we know that he was sympathetic to Copernicus' writings. But, rather than adopt them whole heartedly,

 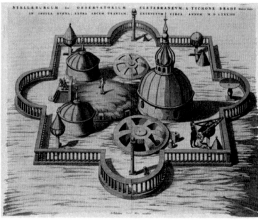

Fig. 1.3 One of the most celebrated astronomers of the sixteenth century, Danishman Tycho Brahe lost his nose in a duel when he was 20. He established a lavish observatory called Uraniborg on the island of Hven near Copenhagen (image of Uraniborg taken from Brahe's *Astronomiae instauratae mechanica* (1598))

Brahe developed his own hybrid model in which all the planets orbited the Sun, but the Sun and the other planets then orbited the Earth (see Fig. 1.4). This model was published in 1588 in *De mundi aetherei recentioribus phaenomenis* [4] ('Concerning the New Phenomena in the Ethereal World'). Unfortunately for Brahe, in this same year his patron King Frederick died after a session of excessive drinking, and his successor King Christian IV was not prepared to continue to fund Tycho's observatory or his lavish lifestyle.

Tycho left Denmark with his family, servants and all the astronomical equipment he could transport and made his way to Prague, where Emperor Rudolph II gave him the position of Imperial Mathematician and gave him money to establish a new observatory in Benatky Castle. This move proved to be fortuitous for Brahe, as it was in Benatky Castle that he met a 29 year old called Johannes Kepler. Kepler came to visit Brahe in January 1600, and started working with the master immediately. They made an ideal team, as Brahe was the diligent observer, whereas Kepler was a diligent mathematician.

Kepler's arrival was just in the nick of time. By October 1601 Brahe was dead, from a bladder or kidney infection after attending a banquet in Prague. We shall never know, but it is possible that his excessive drinking was a contributing factor, but what we do know is that on his deathbed he was preoccupied with his legacy, repeatedly saying 'May I not have lived in vain'. Luckily for him, it is due to what Kepler did with Brahe's observations that Brahe's name has lived on.

Kepler had shown an early interest in mathematics and astronomy. Born into a poor family in Weil der Stadt in what is now southern Germany in 1571, he had a very hard upbringing in a family that suffered constant upheavals due to war, religious strife, a father who was a mercenary and disappeared when Kepler was

1.3 The Man with the Metal Nose

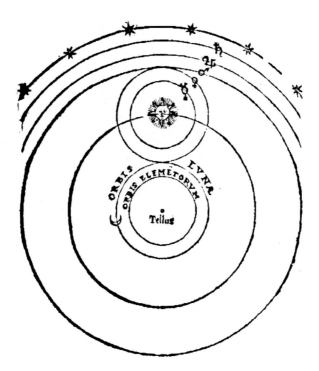

Fig. 1.4 Brahe's hybrid model kept the Earth at the centre of the Universe, with the Sun and other planets orbiting it, but with Mercury and Venus also orbiting the Sun (image from a drawing by Valentin Naboth in *Primae de coelo et terra institutiones* (1573))

only 5 years old; and a mother who was sent away after being accused of witchcraft. But, Kepler found solace in his studies. At only six he observed the great comet of 1577, and at the age of nine he observed a Lunar eclipse. By the age of twenty five he had taught himself enough astronomy to write *Mysterium Cosmographicum* [5] ('Cosmic Mysteries'), which defended Copernicus' De revolutionibus. He was convinced of the correctness of the Sun-centred model, and was determined to find what in Copernicus' model made it less accurate than Ptolemy's geocentric model.

Although Brahe's death was clearly unfortunate for Brahe, really it was the break Kepler needed. Because of Kepler's humble background compared to Brahe's own noble one, it is unlikely that Brahe would have ever properly collaborated with Kepler or treated as anything like an equal. But with Brahe gone, Kepler inherited all of his observing notes, and set about trying to find a model of the Universe that would fit them (Fig. 1.5).

Kepler thought he would have the problem solved in a week or two. In fact, it took him 8 years! His calculations filled nine hundred pages, but in one of the most supreme examples of single minded perseverance, he eventually did produce a model which fitted the observations. One cannot over emphasise what an arduous and difficult task this was, but why did it take him so long?

Fig. 1.5 Johannes Kepler discovered that the planets orbit the Sun in ellipses and not circles (image from a 1610 oil painting of Kepler)

The solution Kepler eventually found required him to abandon one of his most cherished beliefs—that the planets moved about the Sun in perfect circles. For years Kepler stuck to this idea, but he could not get good agreement with Brahe's observations. Eventually, Kepler was able to show that the planets did not move in circles but in ellipses. He also showed that they changed their speed as they orbited the Sun, and that the Sun was not at the centre of the ellipses that the planets followed. Each one of these three findings flew in the face of not just what Copernicus had proposed in his Heliocentric model, but even in the face of Ptolemy's Earth-centred model. In Ptolemy's model the celestial bodies moved in perfect circles at a constant speed along each deferent with the Earth at the geometric centre of these concentric circles.

Whereas most people are familiar with circles, ellipses are a little less familiar, so let me cover the basics necessary to understand what Kepler had discovered. As you may recall from your school days, to draw a circle you can wrap a loop of string with a pencil around a drawing pin, stick the drawing pin into a piece of card, and keeping the loop of string tight you move the pencil around the pin. An ellipse can be produced in essentially the same way, the only difference is that instead of wrapping

1.3 The Man with the Metal Nose

the loop of string around one drawing pin, you wrap it around two of them. The long axis of an ellipse is called the *major axis*, the short axis is the *minor axis*. Half of the length of the major axis is referred to as the *semi-major axis*.

A circle is, in fact, a special case of an ellipse when the two foci are in the same place. The further apart the foci are, the more elongated (elliptical) the ellipse becomes (see Fig. 1.7). This can be quantified as the ratio of the minor axis b to that of the major axis a. However, just to be perverse, mathematicians prefer to talk of the *eccentricity*, denoted by the letter e, which is defined as $e = (1 - b/a)$. If $e = 0$ we have a circle (as the minor axis and the major axis have the same length). If $e = 1$ we would have a straight line, as the minor axis b would have zero length.

Kepler published his findings about the orbit of Mars in 1609 in a book entitled *Astronomia Nova* [6]. In this book he stated two principles of planetary motion, namely

1. That the planets orbit the Sun in ellipses, with the Sun at one of the foci of the ellipse.
2. That the speed of a planet changes in its orbit, sweeping out equal areas in equal times.

Figure 1.6 shows the first of these principles, but what about the second principle? Although it sounds rather complicated, it is in fact not too difficult when one sees a diagram of what it means (see Fig. 1.8). The shaded areas labelled A in Fig. 1.8 all have the same area, but as you can see the distance the planet moves in a time t differs in the three places. To put this another way, when the planet is closer to the Sun (near *perihelion*), it will travel faster, and when it is farthest from the Sun (at *aphelion*) it will travel the slowest. These two principles are now known as Kepler's first and second law of planetary motion.

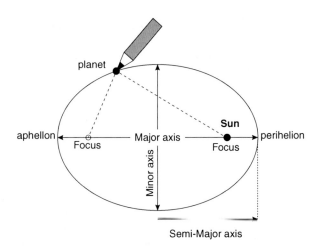

Fig. 1.6 Kepler's 1st law states that the planets orbit the Sun with the Sun lying at one focus of the ellipse

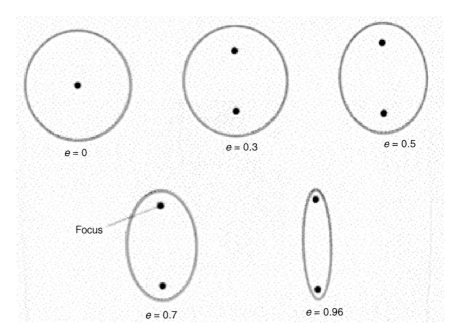

Fig. 1.7 The eccentricity e of an ellipse is a measure of how elongated it is compared to a circle

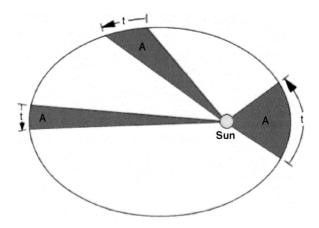

Fig. 1.8 Kepler's second law states that the three *shaded areas* are all the same, meaning that the planet travels quicker when it is near the Sun and slower when it is further from the Sun

Kepler then continued working for another 10 years on various projects. One with which he became obsessed was an idea called 'the music of the spheres', a mistaken belief that the motions of the planets produced musical tones. This was based on an idea Pythagoras had, but Kepler took the idea much further, spending far too much time on such a crazy idea. He also wrote one of the first works of science fiction, a story called *Somnium* [7] ('The Dream') in which a group of adventurers travel to the Moon.

In 1619 Kepler published a third principle of planetary motion in *Harmonices Mundi* ('The Harmony of the World') [8], which we now call his third law. This stated that the square of the orbital period of a planet is proportional to the cube of the semi-major axis of its orbit. This sounds very complicated, but what it basically means is that the more distant planets actually move more slowly than the planets nearer the Sun, in addition to having to travel further in their path. This law actually came out of his work on the music of the spheres, but is the only useful thing to come from that work.

Kepler was also busy compiling data for what became known as the *Rudolphine Tables* [9], which were finally published in 1627, although Kepler had them ready by 1623. The 4 year delay was due to legal wranglings with Brahe's heirs over using his observations. The volume consisted of a star catalogue and tables publishing the positions of the planets many years into the future. The catalogue contained the position of 1,006 stars measured by Brahe, and directions and tables for calculating the positions of the planets. The tables had a level of accuracy never seen before, and amongst other things predicted a transit of Venus in 1631 and a transit of Mercury in the same year.

1.4 Dancing Moons and Crescent Planets

Whilst Kepler was publishing his *Astronomia Nova*, an Italian by the name of Galileo Galilei was exploring the sky using the newly invented telescope. Born in Pisa in 1564, Galileo had started training in medicine, but in 1581 whilst bored sitting in the Cathedral in Pisa listening to a sermon, Galileo was looking at the chandeliers hanging from the ceiling which were moving back and forth due to gusts of wind. He noticed that how long a chandelier took to swing back and forth (its *period*) did not seem to depend on the size (*amplitude*) of the swing. This single observation, and his investigation of it, changed the course of his life.

He persuaded his father to allow him to abandon his medical studies and was allowed to switch to studying mathematics and 'natural philosophy' (as physics was known at the time). In 1589 he was appointed to the Chair of Mathematics at Pisa University, and in 1592 he moved to the University of Padua where he taught geometry, mechanics and astronomy. In 1609, Galileo heard about the invention of an optical instrument that allowed distant objects to appear larger—what we now call a telescope. We don't know who invented the telescope, but in 1608 Dutchman Hans Lippershey tried to patent the idea in the Netherlands. Galileo heard about descriptions of this instrument, and in 1609 set about building his own. The result was a small 2.5 cm aperture telescope with a magnification of about three times (Fig. 1.9).

Galileo quickly set about observing the Moon with his newly fashioned instrument. In late 1609 he made sketches of the Moon, and noticed that its surface was irregular. Looking at the dividing line between the light and dark parts of the Moon, the *terminator*, he correctly deduced that the irregular appearance of this line when

Fig. 1.9 Galileo with the telescope that he built to observe the moons of Jupiter and the phases of Venus (image of Galileo from *Popular Science Monthly Volume 78* (1911). Image of Galileo's telescope by the author)

seen through his telescope was due to changes in elevation on the surface of the Moon—the Moon's surface was not smooth as had been thought for millennia.

As well as being a first rate scientist, Galileo was also a first rate self-publicist, which is why today most people think that he both invented the telescope and was the first person to observe the Moon through one. He was neither. The first person we know of to observe the Moon through a telescope was the English astronomer and mathematician Thomas Harriot, who observed the Moon on the night of the 26th of July 1609, several months before Galileo's first observations. But hardly anyone has heard of Harriot. Indeed, working in West Wales using a telescope made for them by Harriot, Sir William Lower and John Prydderch also used a telescope to observe the Moon in late 1609 and early 1610. Lower even remarked on the rough appearance of the Moon's surface, stating in a letter to Harriot dated the 6th of February 1610 [10] that the surface appeared irregular and

> like a tart that my cooke made me last weeke?

This letter is dated a few weeks before Galileo published his own observations of the Moon, which he did in *Sidereus Nuncius* [11] ('Starry Messenger'), published in March 1610 and summarising Galileo's initial observations with his new instrument (Fig. 1.10).

More profound to our understanding of the heavens were Galileo's observations of Jupiter and Venus. In January 1610 he pointed his telescope towards Jupiter and saw four background stars. Over the course of the following 2 weeks he observed these background stars as they followed Jupiter across the sky, appearing to dance

1.4 Dancing Moons and Crescent Planets 13

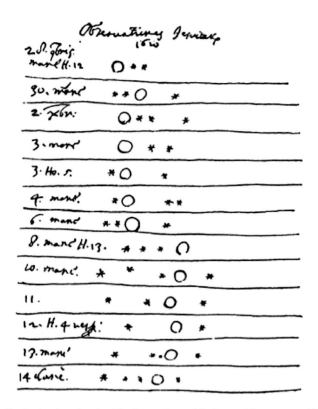

Fig. 1.10 Galileo's sketches showing his observations of Jupiter and the moons orbiting it (image from one of Galileo's observing logs)

around it (see Fig. 1.10). What he had in fact discovered were four moons which were orbiting Jupiter, the first ever evidence that not everything orbited the Earth.

In September of 1611, Galileo started observing Venus. He found that Venus went through phases, something which could not be seen with the naked eye. Although Venus had been expected to show phases, Galileo observed something very important. He saw Venus go through *all phases*, from crescent to full. In addition, through his telescope he saw that Venus showed crescent phases when it also appeared large, but when full it appeared smaller (see Fig. 1.11).

To understand the significance of these observations we need to look at Fig. 1.12. In the Geocentric model, Venus can only show crescent phases at it and the Sun orbit the Earth. However, in the Heliocentric model Venus can exhibit all phases, just as Galileo observed. Additionally, the Heliocentric model naturally explains why Venus appears smaller when full and larger when crescent.

It cannot be overstated how important these observations were. With these observations Galileo had essentially found the proof of the Heliocentric model, showing that the Geocentric model just could not fit the observations. His support of the Copernican model quickly got him into trouble with the Catholic Church.

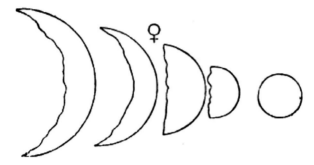

Fig. 1.11 Galileo's sketches of the phases of Venus. Not only did Venus show all phases, but appeared larger when crescent and smaller when full (image from Galileo's *Il Saggiatore* (1623))

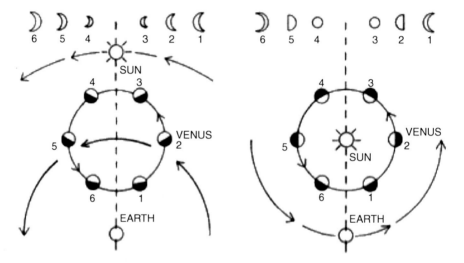

Fig. 1.12 In the Geocentric model, Venus moved on an epicycle as it and the Sun orbited the Earth. In such a model, Venus could only ever exhibit crescent phases as shown on the *left*. Galileo saw Venus go through all phases, from crescent to full and back to crescent. Additionally, its size when full was smaller than when crescent. This is naturally and simply explained if Venus orbits the Sun inside of Earth's orbit of the Sun

In March 1615 the priest Tommaso Caccini went to the offices of the Inquisition in Rome to report to them Galileo's support of a Sun-centred Universe. The Inquisition looked into the matter, and on the 24th of February 1616 they issued the unanimous verdict that the idea that the Sun is stationary is

> foolish and absurd in philosophy, and formally heretical since it explicitly contradicts in many places the sense of Holy Scripture?

Galileo was informed of the verdict the following day, and ordered by Cardinal Bellarmine to stop supporting the Copernican model. Copernicus' *De revolutionises* was suspended 'until corrected', then on the 5th of March the Church banned all books supporting the Copernican system, calling it "the false Pythagorean doctrine, altogether contrary to Holy Scripture". Galileo initially obeyed this Papal decree, but when Pope Paul V died in 1621, he was replaced by Cardinal Barberini, who became Pope Urban VIII, who was a friend and admirer of Galileo. Urban had actually spoken against the condemnation of Galileo in 1616.

Urban encouraged Galileo to write a book giving arguments for and against the Heliocentric model, but Galileo was told to make sure it was impartial. He even asked that the book contain Urban's own views on the matter. Finally, in 1632, with the approval of both the Inquisition and the Pope, Galileo published *Dialogo dei due massimi sistemi del mondo* [12] ('Dialogue Concerning the Two Chief World Systems'). The book had two characters, each arguing the case of the Copernican (Heliocentric) and Aristotelian (Geocentric) models respectively, and a third character acting as an impartial adjudicator.

The arguments for the Copernican model were put forward by a character called Salviati, and those for the Aristotelian model by Simplicio, with Sagredo being the impartial adjudicator. The book portrays Simplicio as an intellectually inept fool, and clearly the name was also meant to get across the message of his limited understanding. Simplicio's arguments for a Geocentric model are systematically refuted and ridiculed by Salviati and Sagredo. In brief, although the book did present both sides of the argument, any reader was left in no doubt which theory Galileo supported.

Soon after the book was published, in September 1632, Galileo was ordered to come to Rome to stand trial. He arrived in February 1633, and in June was found guilty of heresy. He was sentenced to formal imprisonment, but the following day this was changed to house arrest. Dialogue was banned, and Galileo died 9 years later, having gone blind in 1638.

1.5 On the Shoulders of Giants

In the same year that Galileo died in Florence, far away in rural Lincolnshire in England, a baby was born who was to become the towering figure of physics for the next 250 years—Isaac Newton. If Galileo laid the foundations for what we consider to be modern physics and astronomy, it was Newton who built the house. His book *Philosophiae Naturalis Principia Mathematica* [13] ('Mathematical Principles of Natural Philosophy'—usually known as 'The Principia'), published in 1687, is one of the most important publications in the history of science, and laid the blueprint for much of the development of our understanding of the Universe over the next 250 years. I have been fortunate enough to see Newton's original copy of *The Principia*, with his own annotations in the margins. It is on display in the Wren Library of Trinity College, Cambridge. A high quality scanned version of this copy is available at http://tinyurl.com/lzbjob2.

Fig. 1.13 Isaac Newton, born in 1642, towers over the world of Physics. His theories and ideas shaped the development of our understanding of the Universe for the next 225 years

Newton's father, also called Isaac Newton, was a prosperous farmer. Sadly he died 3 months before his son was born, and when his mother re-married, Newton was sent to live with his grandparents. It seems that this experience led to his becoming a particularly loveless and vindictive individual, an aspect of his complex personality we will return to soon. Newton attended The King School in Grantham from twelve until 17 years of age, and in 1661, at the age of eighteen, he went to Trinity College Cambridge. Although by all accounts an unremarkable student, he obtained his B.A. degree in the summer of 1665, and soon after had to leave Cambridge as the University was shut as a precaution against the Great Plague. He returned to his family farm in Lincolnshire where his true genius quickly emerged. He worked on three important pieces of work; his theory of calculus, work on optics, and his ideas on the laws of gravitation (Fig. 1.13).

His work on calculus is a good illustration of his genius, but also of his vindictiveness. Newton realised that to properly understand quantities which changed in an irregular fashion the usual tools of mathematics were not sufficient, so he set about developing what became an entirely new branch of mathematics which we now call calculus. His early work on this subject is in a manuscript dated October 1666, but Newton did nothing about publishing it formally until nearly 30 years later, in 1693. Meanwhile, in 1684, the German mathematician Gottfried Wilhelm Leibniz published his own theory of calculus [14].

Newton assumed that Leibniz had stolen the idea of calculus from him, but modern scholars agree that Leibniz had developed it independently without knowledge of Newton's work on the subject. However, being convinced that Leibniz had stolen his idea, Newton started a personal vendetta against him. In 1699 Newton and fellow members of the Royal Society, which Newton had co-founded in 1662, accused Leibniz of plagiarism. By 1711 the dispute escalated with the Royal Society stating in a study that it was Newton who was the discoverer of calculus and that Leibniz was a fraud who had stolen the idea. Later it was revealed that Newton had been the main author of this study, and had written the concluding remarks on Leibniz himself. But it was a bitter controversy that raged between them until Leibniz's death in 1716.

One of the phrases for which Newton is most famous is

If I have seen further it is by standing on the shoulders of giants

At first sight this seems like a modest statement acknowledging his debt to his scientific predecessors. But, in fact, it was a barb aimed at another man Newton hated with a vengeance, Robert Hooke, a fellow co-founder of the Royal Society. Hooke, most famous for being the first person to describe in detail the structure of a cell, was a hunchback, and Newton deliberately wanted to belittle Hooke's contributions to science, suggesting that Hooke was neither physically nor intellectually a giant.

Despite this nasty personality, it is difficult to imagine how little mathematics and physics would have progressed in the 1600 and 1700s without Newton's genius. The year of 1666 is often referred to as Newton's *'annus mirablus'*, as not only did he write his first manuscript about calculus, but he also came up with one of his other great ideas, the law of universal gravitation. Whether the story of his insight into the nature of gravity being triggered by seeing an apple falling from a tree is true or not, Newton was the first person to understand the most fundamental law in Nature, and by doing so was able to explain why objects fall to the Earth, why the Moon is held in its orbit about the Earth, and why the planets orbit in ellipses about the Sun.

In short, he laid out in The Principia, an essentially complete explanation of the motions of moving bodies under forces, and a mathematical explanation of the only force then known, the force of gravity. Although his ideas would be superseded by Einstein's theories of Special and General Relativity in the early twentieth century, we still use Newton's laws of motion and gravity for nearly every day-to-day application such as designing the brakes of a car or sending a space probe to land on Mars.

1.6 The Relative Size of the Solar System

Careful observations and some simple mathematics had allowed astronomers to determine by the end of the 1600s the sizes of the orbits of the planets relative to the distance from the Earth to the Sun. For example, by measuring the angle between Venus and the Sun when Venus is furthest East of the Sun (what is called *maximum eastern elongation*), one can determine with very simple trigonometry that the size of Venus' orbit is 0.7 of the size of Earth's orbit. Exactly the same method yielded the size of Mercury's orbit as being nearly 0.4 of the size of Earth's orbit.

To determine the relative sizes of the orbits of the so-called *superior planets* (Mars, Jupiter and Saturn were the only ones known at this time) required a slightly more complicated method. The angles between the superior planet when at conjunction and at quadrature needed to be measured. Conjunction is when the superior planet lies directly in the opposite direction to the sky, so that the planet, the Earth and the Sun form a straight line. Quadrature is when the angle from the Sun to the Earth to the superior planet forms a right angle. Then after a little bit of trigonometry the relative size of e.g. Mars' orbit could be determined.

Using these methods, astronomers knew by the end of the 1600s the relative sizes of the planets' orbits compared to the size of Earth's orbit. But, they did not know the actual distances involved. This is a little like looking at a map and seeing that Edinburgh is twice as far from London as Leeds is, but not knowing what the actual distance is from London to Edinburgh. Estimates of the distance from the Earth to the Sun varied wildly, from a few tens of thousands of kilometres to millions of kilometres. Clearly it was an important problem to solve. The man to think of a reliable way to do it was Edmund Halley.

1.7 Edmund Halley, le Gentil, Captain Cook and Sex for Iron Nails

Halley was one of Newton's contemporaries at the Royal Society, best known for having a comet named after him. But, Halley was much more besides. In a long career he was a sea captain, a map maker, the second Astronomer Royal, the professor of Geometry at Oxford University, he invented the deep sea diving bell, and was the person who came up with the idea of how to measure the size of the Solar System. In 1676, at the age of just nineteen, Halley was sent by the Royal Observatory in Greenwich to the South Atlantic island of Saint Helena (where Napoleon Bonaparte would later be imprisoned) to map the stars of the Southern Skies. Whilst there, on the 7th of November 1677, he observed a transit of Mercury, when Mercury is seen to move across the disk of the Sun.

He soon realised that if one were to observe a transit of Mercury, or the more rare transit of Venus, simultaneously from two different locations on Earth then due to the effect of parallax the planet would follow different paths across the Sun's disk as seen from the two different locations. By knowing the distance between the two observing locations, and measuring the angular distance between the two paths on the disk of the Sun, one could use the principle of triangulation to determine the distance from the Earth to the Sun (Fig. 1.14).

In 1716, nearly 25 years later, Halley presented this theory to the Royal Society [15], by which time he had realised that the distance from the Earth to Mercury would be too large to use this triangulation technique, but a transit of Venus should work (Fig. 1.16). Transits of Venus occur in pairs separated by 8 years, and then a long gap of either 105.5 or 117.5 years until the next pair. A pair of transits were

1.7 Edmund Halley, le Gentil, Captain Cook and Sex for Iron Nails

Fig. 1.14 Edmund Halley was one of the most remarkable scientists of the seventeenth century, if not of any century. He was the second Astronomer Royal, professor of Geometry at Oxford University, and by all accounts was the person who prompted Newton to start writing his *Principia* (image from a portrait by Thomas Murray (1687))

due to occur in 1761 and 1769. Halley knew he would not live to see these transits, but he encouraged his fellow astronomers to take up the challenge to determine the distance from the Earth to the Sun.

The idea Halley had is simple to understand. If a Transit of Venus is observed from two different locations on Earth separated by as much distance as possible in a North-South direction, then Venus will appear to move across a different part of the Sun due to the effect of parallax. By timing the moment when Venus touches the Sun's disk at the start and leaves the Sun's disk at the end of its transit, one could determine the angle of the transit line (the chord in Fig. 1.15) from the line across the diameter of the Sun.

When the June 1761 Transit came, over 120 astronomers from some eight nations were dispatched to make measurements. As luck would have it, the transit was only visible in its entirety from central and eastern Asia (including the Indian Ocean) and only partially visible from Europe and Africa. This meant the astronomers, mainly from European countries, had to set off on long voyages by either land or sea. Most of the astronomers were from Britain and France, and to add to their woes the two countries had started what would become known as the '7-year war' in 1756.

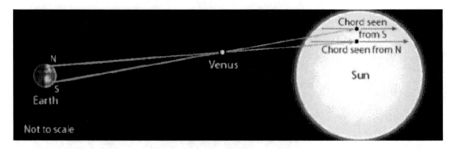

Fig. 1.15 If one observes a Transit of Venus from two different locations separated in a North-South direction, the planet will appear to move across a different part of the Sun due to the effect of parallax

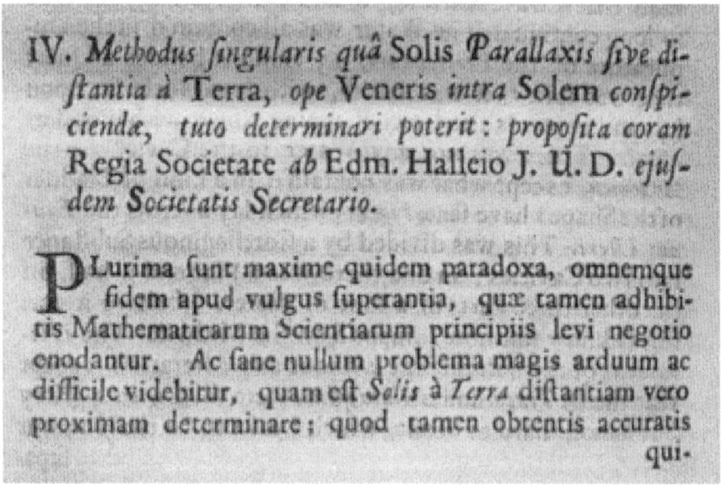

> IV. *Methodus singularis quâ* Solis *Parallaxis sive distantia à* Terra, *ope* Veneris *intra* Solem *conspiciendæ, tuto determinari poterit: propofita coram* Regia Societate *ab* Edm. Halleio J. U. D. *ejufdem Societatis Secretario.*
>
> Plurima funt maxime quidem paradoxa, omnemque fidem apud vulgus superantia, quæ tamen adhibitis Mathematicarum Scientiarum principiis levi negotio enodantur. Ac sane nullum problema magis arduum ac difficile videbitur, quam est *Solis* à *Terra* distantiam vero proximam determinare; quod tamen obtentis accuratis qui-

Fig. 1.16 Halley's original 1716 paper on using a Transit of Venus to determine to distance from the Earth to the Sun

Although the astronomers from the two countries were willing to collaborate, their respective countries certainly were not. Several of the expeditions by both sides were affected by the war. Typical of this was the expedition undertaken by Charles Mason and Jeremiah Dixon. At the time, Mason was the assistant to the Greenwich Observatory astronomer Neville Maskelyne, and he was originally intended to accompany Maskelyne on an expedition to Saint Helena to observe the Transit. But, at the last minute the Royal Society decided that Mason could lead his own expedition to Bencoolen, which was a British possession in modern-day Sumatra. Mason chose Dixon to accompany him, and in December 1960 they set sail from Plymouth on the British warship *HMS Seahorse*.

Within hours of leaving port they were fired upon by the French frigate *Le Grand*, and returned to port with eleven dead and thirty seven wounded. Reluctant to continue, Mason sent a letter to the Royal Society stating that they were unwilling to undertake the expedition because of the dangers. The Royal Society threatened to prosecute, so reluctantly Mason and Dixon set sail again on February the 3rd on a repaired Seahorse. They took 3 months to reach the tip of Africa, whereupon they heard that Bencoolen had fallen into French hands. They decided to stay where they were, and observed the Transit on June the 6th from Cape Town in modern-day South Africa. They obtained excellent data, whereas Mason's boss Maskelyne was entirely clouded out on Saint Helena.

Mason and Dixon clearly got on, as a few years later they were dispatched by the British Crown to settle an on-going land dispute between Lord Baltimore of Maryland and William Penn of Pennsylvania in the Colonies. They surveyed a line which laid down the border between the two territories, a line which we now call the 'Mason-Dixon line'. It became better known during the American Civil war as the dividing line between the 'slave south' and the 'free north', and is the origin of the term 'dixie' for the southern states of the USA.

But, if Mason and Dixon did not have the best of luck, their misfortune pales into insignificance compared to that suffered by a French nobleman with the unlikely name of *Guillaume Joseph Hyacinthe Jean-Baptiste le Gentil de la Galasière*, whom we shall from now on refer to by the slightly shorter name 'le Gentil'. France sent more astronomers to observe the 1761 Transit than any other country, and le Gentil was one of them. He sailed from Brest in north-western France on March the 26th 1760 on board the fifty-gun *Le Berryer*, bound for Pondicherry in eastern India (just to the south of modern-day Chennai), which at the time was a French possession.

After 3 months at sea, his ship arrived in Isle de France (modern-day Mauritius) in July 1760, nearly a year before the transit. There le Gentil learnt that Kirakel, a French settlement just to the south of Pondicherry, had just been captured by the British and that Pondicherry itself was under siege. A French fleet assembled at the Isle de France with the intention of relieving Pondicherry and transporting le Gentil there to observe the transit. But, in January 1761 this fleet was wrecked by a hurricane, and le Gentil decided to abandon going to Pondicherry and wrote to the French Academy of Science that he was gong to Batavia instead. Then, he went down with dysentery.

It was March before le Gentil was well enough to travel, by which time a ship called *la Sylphide* was due to sail to Pondicherry to help defend it against the advancing British. Le Gentil decided to go to his original destination, so they set sail. However, it was monsoon season, and the ship was blown off course. They wandered around the Indian Ocean off the coast of Africa and the Arabian peninsula, and then on the 24th of May, they learnt from other French ships that Pondicherry had fallen into the hands of the British. The captain of the ship made the decision to return to Isle de France. When the day of the Transit came, Le Gentil found himself on the rolling deck of la Sylphide in the middle of the Indian Ocean, unable to obtain any readings of the Transit. When he got back to Isle de France he made the remarkable decision to hang around for the next transit in 8 years' time!

When the results from the 1761 Transit were compiled they gave very mixed results, with the calculation of the distance of the Earth from the Sun ranging from 77 million miles to 97 million miles. Although astronomers had hoped for better, this was still an improvement on the estimates made by astronomers like Christian Huygens (100 million miles) and Giovanni Cassini (86 million miles), estimates which were based entirely on guesswork and assumptions.

One of the reasons for the large variation in the results is that astronomers had underestimated how difficult it would be to time exactly when the Transit started and finished, a measurement which was hampered by something dubbed the 'black drop effect' (see Fig. 1.17), where the dark disk of Venus appeared to leak towards the edge of the Sun, making timing the exact contact time very difficult.

Luckily there was another Transit in June 1769 which gave astronomers a second chance to get more accurate readings. Knowing that this would be the last transit until the December 1874 transit, an even greater effort was undertaken. 151 observing teams were sent to 77 sites around the World. Again, the majority of these teams were from European countries, and again the Transit of 1769 was only visible on the other side of the World. Perhaps the most famous team was the one led by

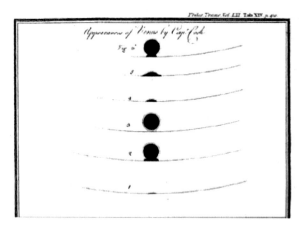

Fig. 1.17 The 'black drop effect' made it difficult to time the exact contact time between Venus and the limb of the Sun (this series of drawings is from Cook's observations in Tahiti in 1769)

James Cook, who was dispatched by the Royal Society to the South Pacific island of Tahiti.

Cook was a remarkable seaman, the first ever captain in the Royal Navy to have risen up through the ranks and to not have come in as a commissioned officer. He started his working life as a grocer's apprentice, but soon after left to become an apprentice on a coal ship called *Freelove* taking coal from Newcastle to London. He joined the Royal Navy at the start of the 7 year war, and distinguished himself during the siege of Quebec in 1759 as an excellent navigator and marine surveyor.

The Royal Society's transit committee wanted to observe the transit from as close as possible to its 'Halleyan point', which is where the transit would appear the shortest. The Halleyan point was in the Southern Pacific, at that time a largely uncharted body of water. The committee debated between sending an expedition to Marquess or Tonga, but neither island had been seen since Europeans had first sighted them. In May 1768 Samuel Wallis suggested an alternative—Tahiti. He had just returned after a round-the-world voyage in the *HMS Dolphin*, and having come across Tahiti by accident he had accurately determined its longitude, thus making it much easier to find again.

During their time in Tahiti, Wallis had noticed that the Tahitian women had a particular fondness for iron, so much so that they were willing to exchange sexual favours for it. The most readily available source of iron were the nails which held the wooden-hulled Dolphin together, and so eager were the ship's crew to satiate the women's desires that they stripped as many nails as they could from the ship. The ship nearly fell apart before Wallis realised what was going on and put a stop to his crew's plundering.

A ship called *The Earl of Pembroke* was bought by the Royal Navy and refitted for the voyage to Tahiti. It was renamed *HMS Endeavour*, and Cook was promoted to the rank of Lieutenant and chosen to captain the ship and her crew. They set sail from Plymouth on the 26th of August 1768. Having been a common sailor himself, Cook knew there was little he could do to change the behaviour of his men should the Tahitian women make the same offer of sex for iron that they had to Wallis' crew. Not wanting his crew to start tugging nails out of Endeavour, he stashed a secret supply of spare nails in a large barrel which he hid away in his cabin.

Cook's party included the astronomer Charles Green, who was the assistant to the now Astronomer Royal Maskelyne, and botanist Joseph Banks. Endeavour sailed across the Atlantic, rounded the southern tip of South America and arrived in Tahiti on the 13th of April 1769. The party set about building a fortified observatory at what is now called Point Venus, in the north of the island, to make their observations of the Transit, which was on the 3rd of June in Tahiti (the 4th of June for places further west). On the day of the Transit they were blessed with clear blue skies, and obtained excellent data of the timings of Venus' passage across the Sun.

Once the Transit was over, Cook opened some sealed orders which had been given to him by the Royal Navy. These were instructions to go in search of *'Terra Australis Incognita'*, the 'unknown southern lands'. At the time, geographers believed that there should be a balance between the land masses to be found in the northern and southern hemispheres, and so thought there was a large, undiscovered land-mass somewhere in the southern oceans. In September 1769 the Endeavour arrived in New Zealand, being the second Europeans to visit there after Dutch explorer Abel Tasman in 1642. There, Cook spent some 6 months mapping the coast and showing, for the first time to Europeans, that the land was made up of two separate islands. In the northern part of the North island they observed the November 1769 Transit of Mercury, from a place now known as Mercury Bay.

On the 31st of March 1770 Cook recorded in his Journal that they would return home in a direction which would take them towards the land of 'New Holland', as Australia was then known. The Dutch had explored the west coast, but the remainder of the land mass was unchartered. Originally Cook intended to sail to Tasmania, or 'Van Diemen's Land' as it was known at the time, which had also been sighted by Tasman. Cook wanted to see whether Van Diemen's Land formed part of the proposed Terra Australis Incognita. However, they sailed too far north, and missed it. Instead, on the 20th of April, they sighted land, which turned out to part of the Australian landmass. They landed at what is now known as Botany Bay on the 29th of April, and claimed the land for the British Crown.

Meanwhile, after his failure to observe the 1761 Transit, le Gentil (Fig. 1.18) had bided his time waiting for the 1769 Transit. He spent some time studying the flora and fauna of Madagascar, and even mapped its east coast, and then decided he would observe the 1769 Transit from Manila in the Philippines. But, when he got

Fig. 1.18 Guillaume le Gentil, who was gone from is home in Paris for "11 years, 6 months and 13 days" in an attempt to observe the 1761 and 1769 Transits of Venus

1.7 Edmund Halley, le Gentil, Captain Cook and Sex for Iron Nails

there he found the Spanish authorities were hostile to his presence, so he headed to Pondicherry, which had been restored to the French in a peace treaty signed with the British in 1763.

He arrived in March 1768, and built an Observatory in the grounds of the abandoned British fort, and waited patiently for the Transit on the 4th of June the following year. In the weeks leading up to the Transit, le Gentil enjoyed clear skies day after day and night after night. But, on the night before the Transit he was kept awake by the sound of the wind picking up, and awoke on the morning of the 4th of June to find the sky covered in thick dark storm clouds. le Gentil saw nothing, not a single glimpse of Venus moving across the disk of the Sun. He wrote in his journal

> I was more than two weeks in a singular dejection and almost did not have the courage to take up my pen to continue my journal.

Le Gentil's misfortune had not ended. Utterly forlorn, he waited for a ship to take him home to France, but whilst waiting he contracted dysentery again, and this time it nearly killed him. After a few weeks in bed he regained his strength enough to board a ship bound for Isle de France (Mauritius), and from there he boarded a ship bound for France. However, off the coast of the Cape of Good Hope, this ship was damaged in a hurricane and had to return to Isle de France.

The next ship out was a Spanish troop ship which was bound for Cadiz, but so desperate was le Gentil to get home that he boarded it and made the last part of his voyage overland from Cadiz to Paris. He got back to Paris "11 years, 6 months, 13 days" (to quote his Journal) after he had left in the March of 1760, only to find that his relatives had him declared dead and his entire estate had been divided up amongst his relatives! His luck did, however, finally change. The French Academy created a special university Chair for him, he wrote two volumes of his memoirs, and met and married a wealthy heiress Mme. Potier with whom he had a daughter. He died peacefully at home on October the 22nd, 1792, just before the French reign of terror began.

When the results of the observations of the 1769 Transit were put together they yielded much more consistent results. In 1771, Thomas Hornsby, Professor of Astronomy at Oxford University, presented his analysis of five separate observations to the Royal Society [16]. He used observations from Vårdö in Northern Scandinavia, the Kola Peninsula (between the Arctic Ocean and the White Sea), Hudson's Bay in Canada, Baja California and Cook's observations from Tahiti. The average distance from the Earth to the Sun given by these five sets of observations was 93,726,900 miles. This is within 1 % of the currently accepted value of 92,957,133 miles which is determined by bouncing RADAR off of the surface of Venus. Finally, astronomers knew the scale of the Solar System!

1.8 Do the Stars Move?

The next step astronomers needed to make was to determine the distance to the stars. Remember, one of the objections used in arguing against the heliocentric model was that our changing perspective as we move around the Sun should lead to stars appearing to shift their positions in the sky—so called *stellar parallax*. No such parallax had ever been observed, and for many decades this was used by people who wanted to cling on to the belief that the Earth did not move in space and was at the centre of the Universe. In 1729 James Bradley, the third Astronomer Royal and the one who succeeded Edmund Halley in the position, was the first to properly attempt to measure stellar parallax. But, he was unable to detect any shift of in stars' position.

The first successful stellar parallax measurement was made by German mathematician and astronomer Friedrich Bessel [17]. In 1838 he measured a tiny motion of the star 61 Cygni, and when I say tiny I mean tiny. He found that the star exhibited a parallax of $0.3136''$ (0.3136 arc seconds, remember there are 60 arc seconds in one arc minute, and 60 arc minutes in one degree so an arc second is 1/3600th of a degree!). An object which is 44 cm in diameter would appear to have this angular size if it were 5 km away! The stellar parallax measured by Bessel translated into a distance for the star of 3.19 parsecs or 10.40 light years, which is pretty close to the currently accepted value of 3.48 parsecs or 11.36 light years.

So fiendishly small were the parallax angles that, by the end of the nineteenth century, only about sixty stars had their distances determined using this method. It is the only method astronomers have for directly measuring distances beyond the Solar System, and yet it was only able to yield the distances to a handful of the nearest stars. Most stars, as they showed no sign of a measurable parallax, were presumed to be much further away. How were astronomers going to determine their distances?

1.9 Mapping the Milky Way and Nebulae

William Herschel, the German-born musician turned astronomer, is most famous for being the first person to discover a planet. In March 1781 from the back garden of his house in Bath, using a 6-in. reflecting telescope that he had built himself and whilst searching for double stars, he spotted an object which looked like a non-stellar disk. Originally, Herschel thought it was a comet or possibly a star, but observations over subsequent weeks showed that it was moving against the background stars. Herschel realised he had found a Solar System object, and he also realised that it was orbiting beyond the orbit of Saturn, the most distant planet known at that time. He called the new planet *Georgium sidus* ('Georgian star'), in honour of King George III, which he hoped would win him favour with the King. The name did not stick, and

1.9 Mapping the Milky Way and Nebulae

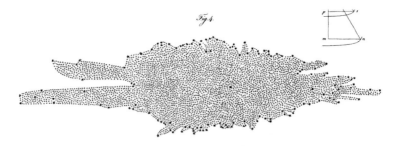

Fig. 1.19 Herschel's map of the Milky Way which came from counting the stars in over 600 different locations and assuming they all had the same intrinsic brightness (image from a paper entitled *On the Construction of the Heavens* by Herschel published in *Philosophical Transactions of the Royal Society of London, Vol. 75* (1785))

eventually the name Uranus was given to the new planet, following the convention of naming planets after Roman gods.

The following year, in 1782, Herschel was appointed 'The King's Astronomer', a new position created by King George III, and he and his sister Caroline moved from Bath to Datchet, a village which is on the outskirts of modern-day Slough to the west of London. Herschel built two larger telescopes, a 20-ft focal length reflector with a 12-in. aperture and an 18.7-in. aperture telescope (called the '40-ft'). He also continued to use the 6-in. reflector with which he had discovered Uranus. At his new location, Herschel embarked on two major observing projects. The first was to try to determine the layout of stars in the Milky Way. To do this he catalogued the position of every star he could see in over six hundred different parts of the sky and, assuming they all had the same intrinsic brightness, created a map of our stellar system (see Fig. 1.19).

He also set about compiling a catalogue of 'deep sky' (non-stellar) objects. Eventually he would discover over 2,400 objects which he defined as 'nebulae' (an object which appeared diffuse or extended). He published his discoveries in three catalogues, 'Catalogue of One Thousand New Nebulae and Clusters of Stars' (1786), 'Catalogue of a Second Thousand New Nebulae and Clusters of Stars' (1789) and 'Catalogue of 500 New Nebulae, nebulous Stars, planetary Nebulae, and Clusters of Stars' (1802) [18–20].

In these catalogues he arranged his discoveries into eight classes: (i) bright nebulae, (ii) faint nebulae, (iii) very faint nebulae, (iv) planetary nebulae, (v) very large nebulae, (vi) very compressed and rich clusters of stars, (vii) compressed clusters of small and large (faint and bright) stars, and (viii) coarsely scattered clusters of stars. Herschel's sister Caroline discovered an additional eleven objects, and his son John Herschel discovered a further 1754, which he published along with his father's and aunt's in 'Catalogue of Nebulae and Clusters of Stars' in 1864 [21]. This catalogue was later edited by John Dryer, with additions from many other nineteenth Century astronomers, and published in 1888 as the 'New General Catalogue' [22], which is from where the 'NGC' designation for so many nebulae comes.

1.10 Stellar Fingerprints

In 1814–1815, Joseph von Fraunhofer noticed that the Sun's spectrum contained a number of dark lines, superimposed on the familiar rainbow colours which Newton had discovered in 1670–1672. Fraunhofer published these results in 1817 [23] (see Fig. 1.20). In 1859 it was shown by Gustav Kirchhoff and Robert Bunsen (after whom the 'bunsen burner' is named) that bright spectral lines could be produced by sprinkling different salts into a burning flame. Over the next several decades, it was found that different chemical elements had different spectral lines.

The lines could either be seen as bright lines (so-called *emission spectra*), or as dark lines (*absorption spectra*), depending on the experimental configuration. When we look at the spectrum of stars (including the Sun), we usually see dark lines, and by carefully studying the pattern of these lines we can identify the elements in the atmospheres of the stars, which is where the dark lines are being produced. In fact, the element Helium was first discovered in the spectrum of the Sun, before the element had ever been discovered on Earth. Its name comes from *Helios*, the Greek word for the Sun.

The first photograph of a stellar spectrum was made in August 1872 when a rich amateur astronomer by the name of Henry Draper photographed the spectrum of Vega. Draper was trained as a physician, and was the son of John William Draper, who was a professor at New York University. Draper followed his father into medicine, and also became a professor at the same university. In 1867 he married rich socialite Anna Mary Palmer. Although medicine was his profession, his passion had always been astronomy. In the winter of 1839/1840, he took the first known photograph of the Moon. Soon after taking the first photograph of a stellar spectrum in 1872, he resigned his position at New York University so he could concentrate full-time on his astronomy. He took the first known photograph of the

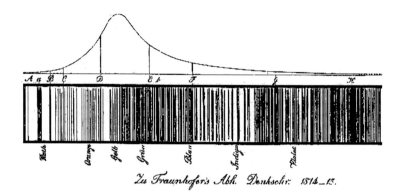

Fig. 1.20 Joseph von Fraunhofer's sketch of dark lines on the spectrum of the Sun, which he discovered in 1814–1815

Orion Nebula in 1880, but sadly died of pleurisy in 1882. By the time of his death he had photographed over one hundred stellar spectra.

In 1885, building on this pioneering work, Edward Pickering the Director of Harvard College Observatory since 1877, began a systematic programme to photograph as many stellar spectra as was possible. In 1886 Draper's widow provided funds for what would become known as the *Henry Draper Catalogue*, which when published between 1918 and 1924 contained the spectra of some 225,300 stars! As part of this work, astronomers started the first systematic classification of stellar spectra, something which led to the 'OBAFGKM system' we still use today.

Key to Pickering's programme was his use of a group of well-educated and highly talented women, who became known as *'Pickering's Harem'*. It was one of these women, Annie Jump Canon, who formalised the OBAFGKM classification system. Another one, Henrietta Leavitt, was to finally provide a way to map our Milky Way.

1.11 Cepheid Variables and Standard Candles

How could astronomers hope to determine the proper structure of the Milky Way if they could only measure the distances to the nearest stars? Assuming, as William Herschel had done, that they all had the same intrinsic brightness was clearly not correct, as the few dozens of nearby stars whose parallax had been measured showed. The key was discovered by accident by Henrietta Leavitt, who was part of Pickering's Harem (Fig. 1.21). She found a remarkably useful property of a class of stars known as *Cepheid variables*.

The first Cepheid variable to be discovered was delta Cephei, by the English astronomer John Goodricke in April 1784. Goodricke noticed that the star's apparent brightness varied regularly with a period of several days. Modern measurements give the change in brightness between its dimmest and brightest as 0.89 magnitudes, corresponding to a factor of nearly 2.5, and the period of this change is 5.37 days. Over the course of the next one hundred or so years, more and more stars were found to show similar behaviour, with periods ranging from a few hours to several tens of days. Probably the best known star which is a Cepheid variable is Polaris, the north star.

Pickering gave Leavitt the task of studying these variable stars, and in 1908 she made a key discovery. Whilst studying stars in a large diffuse nebulae known as the *Magellanic Cloud*, she found thousands of variable stars [24]. She noticed that some varied over a few hours, others over a few days, and others over tens of days. She also noted that the stars which took the longest to vary their brightness appeared to be the brightest of the variable stars at maximum brightness, and the ones which took the least time appeared to be the faintest. But, as these stars were all in the Magellanic Cloud, Leavitt realised that the stars were all at roughly the same distance. This meant that the longer period Cepheids were *intrinsically* brighter than the shorter period ones. This was crucial, as it enabled her to show that a relationship existed between the period of a Cepheid variable's change and

Fig. 1.21 Henrietta Leavitt, working at her desk at the Harvard College Observatory. Leavitt discovered the period-luminosity relationship for Cepheid variable stars, and who also coined the term "standard candle"

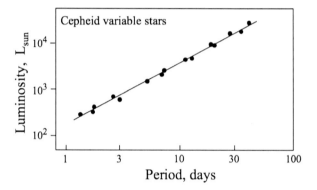

Fig. 1.22 The period-luminosity relationship for Cepheid variable stars, as found by Henrietta Leavitt. Intrinsically brighter stars take longer to vary their brightness than fainter ones. L_{sun} refers to the luminosity of the Sun

the intrinsic brightness of the star when at its brightest. The work was published in 1912 [25], although as was the custom of the time the author was Director Pickering, with Leavitt only cited in the paper. This important relationship is now known as the *period-luminosity relationship*. The relationship is shown in Fig. 1.22, the x-axis is the period i days, and the y-axis is the intrinsic brightness of the Cepheid in units of 'solar luminosity', the luminosity of the Sun.

Leavitt had discovered the first *'standard candle'* in astronomy, a way to determine the distance of a star from its observed brightness. In fact, the term standard candle is one Leavitt herself coined. All that was required was to find a Cepheid variable that was close enough to have its distance measured using the stellar parallax technique, and then the intrinsic brightnesses and hence distances of all other Cepheid variables could be determined from Leavitt's period-luminosity relationship. The Danish astronomer Ejnar Hertzsprung was the first to do this, in 1913 he was able to determine the distance to several Cepheids, allowing this technique to bear fruit. As we shall see in Chap. 2, Cepheids played a crucial role in determining not only our place in the Milky Way galaxy, but in determining the size of our Milky Way, and in showing that our Galaxy was just one of many in the Universe.

References

1. Ptolemy, C.: The Almagest (2nd Century). Available online at e.g. http://www.wilbourhall.org/index.html
2. Copernicus, N.: Commentariolus ("Little Commentary"). The complete Latin text is available at e.g. http://tinyurl.com/k758tne (1514)
3. Copernicus, N.: De revolutionibus orbium coelestium ("On the Revolutions of the Heavenly Spheres"). Available online at e.g. http://ads.harvard.edu/books/1543droc.book (1543)
4. Brahe, T.: De mundi aetherei recentioribus phaenomenis ("Concerning the New Phenomena in the Ethereal World"). Available online at e.g. http://tinyurl.com/pcprasb (1588)
5. Kepler, J.: Mysterium Cosmographicum ("Cosmic Mysteries"). Available online at e.g. http://www.e-rara.ch/doi/10.3931/e-rara-445 (1577)
6. Kepler, J.: Astronomia Nova. Available online at e.g. http://www.e-rara.ch/zut/content/titleinfo/162514 (1609)
7. Kepler, J.: Somnium ("The Dream"). Available online at e.g. https://archive.org/details/operaomniaedidit81kepluoft (1608)
8. Kepler, J.: Harmonices Mundi ("The Harmony of the World") Available online at e.g. https://archive.org/details/ioanniskepplerih00kepl (1619)
9. Kepler, J.: Tabulae Rudolphinae ("Rudolphine Tables"). Available online at e.g. http://tinyurl.com/najp8lx (1627)
10. As quoted in King, H.C.: The History of the Telescope. Charles Griffin and Co., London (1955)
11. Galileo, G.: Sidereus Nuncius ("Starry Messenger"). Available online at e.g. http://tinyurl.com/qhajsdp (1610)
12. Galileo, G.: Dialogo dei due massimi sistemi del mondo ("Dialogue Concerning the Two Chief World Systems"). Available online at e.g. http://tinyurl.com/k9e26ad (1632)
13. Newton, I.: Philosophiae Naturalis Principia Mathematica ("Mathematical Principles of Natural Philosophy"). Available online at e.g. http://tinyurl.com/lzbjob2 (1687)
14. Leibniz, G.: Nova Methodus pro Maximis et Minimis ("New method for maximums and minimums"). Available online at e.g. http://tinyurl.com/nuamsl4 (1684)
15. Halley, E.: Methodus Singularis quâ Solis Parallaxis sive distant à Terra, ope Veneris intra Solem conspiceindae ("A New Method of Determining the Parallax of the Sun, or His Distance from the Earth"). Philos. Trans. R. Soc. **29**, 454–464 (1716). For an English translation see e.g. http://tinyurl.com/o6mp5ss
16. Hornsby, T.: The quantity of the Sun's parallax, as deduced from the observations of the transit of venus, on June 3, 1769. Philos. Trans. R. Soc. **61**, 574–579 (1771)
17. Bessel, F.W.: Bestimmung der Entfernung des 61sten Sterns des Schwans ("Determination of the distance to 61 Cygni"). Astron. Nachr. **16**, 65–96 (1838)
18. Herschel, W.: Catalogue of one thousand new nebulae and clusters of stars. Philos. Trans. R. Soc. **76**, 457–499 (1786)
19. Herschel, W.: Catalogue of a second thousand of new nebulae and clusters of stars; with a few introductory remarks on the construction of the heavens. Philos. Trans. R. Soc. **79**, 212–255 (1789)
20. Herschel, W.: Catalogue of 500 new nebulae, nebulous stars, planetary nebulae, and clusters of stars; with remarks on the construction of the heavens. Philos. Trans. R. Soc. **92**, 477–528 (1802)
21. Herschel, J.F.W.: Catalogue of nebulae and clusters of stars. Philos. Trans. R. Soc. **154**, 1–137 (1864)
22. Dreyer, J.L.E.: A new general catalogue of nebulae and clusters of stars, being the catalogue of the late Sir John F.W. Herschel, Bart., revised, corrected, and enlarged. Mem. R. Astron. Soc. **49**, 1–237 (1888)

23. Fraunhofer, J.: Bestimmung des Brechungs- und des Farben-Zerstreuungs - Vermögens verschiedener Glasarten, in Bezug auf die Vervollkommnung achromatischer Fernröhre ("Determination of the refractive and color-dispersing power of different types of glass, in relation to the improvement of achromatic telescopes"). Ann. Phys. **56**, 264–313 (1817)
24. Leavitt, H.S.: 1777 Variables in the Magellanic Clouds. Annals of the Harvard College Observatory, vol. IV. Observatory, Cambridge (1908)
25. Pickering, E.C.: Periods of 25 variable stars in the small magellanic cloud. Harv. Coll. Obs. Circ. **173**, 1–3 (1912)

Chapter 2
A Universe of Galaxies

In the early 1900s Kapteyn argued that our Sun was at the centre of the Milky Way galaxy, and Slipher found that most "spiral nebulae" were moving away from us. In 1920 Shapley and Curtis debated the scale of the Universe, with Curtis arguing that our Galaxy was just one of many and Shapley arguing that it was the entire Universe. In 1923 Hubble discovered Cepheid variables in the Andromeda "nebula", and showed conclusively that it was far beyond our Milky Way. He went on to show in 1929 that more distant galaxies are moving away from us more quickly, the most natural interpretation of this is that the Universe is expanding.

2.1 Kapteyn's Universe

In August of the first year of my Ph.D., I made my first trip to a research observatory. I was awarded seven nights of observing time on a 1-m telescope in La Palma, the most north-westerly of the Canary Islands off the west coast of North Africa. The telescope is part of the "Isaac Newton Group" of telescopes, which are themselves part of the numerous telescopes at the *Roque de los Muchachos Observatory* on the rim of the caldera of La Palma island's volcano, at an altitude of 2,396 m (7,861 ft). The name of the telescope I was going to use was the *Jacobus Kapteyn Telescope* (JKT), named in honour of the Dutch astronomer. At that time there were no direct flights from the UK to La Palma, the only way to get there was to fly to Tenerife and then take a flight from there onto to La Palma. This wasn't really a hardship for a young Ph.D. student, as it necessitated a few days in Tenerife, where I could sunbathe and enjoy the sights of that far more busy island before I went up to the solitude of the mountain-top observatory.

Once I had landed on La Palma, the taxi from the airport took me up one of the steepest and most spectacular roads I have ever travelled. Apparently the rise from sea level to the Observatory is the biggest change in altitude for the distance travelled of anywhere in the World. The views from the Observatory are spectacular, both in the day and at night. As the Sun sets, one can look across to the East to the island of Tenerife. The peak of Mount Teide on Tenerife rises to just over 3,700 m, considerably higher than the peak I was standing on. This meant that the top of Mount Teide was picked out by the last dying rays of the Sun, even as I on the peak at La Palma had already been plunged into twilight.

Once the Sun had set and darkness had fallen, the view of the heavens was breathtaking. Not only is the Observatory in an exceptionally dark site, but the clarity and steadiness of the air is one of the best in the Northern Hemisphere, reducing the characteristic twinkling of stars to a barely perceptible level. As I was there in August, the band of the Milky Way stretched high overhead for most of the night. I had never seen the Milky Way with such clarity, and I grew up in west Wales, which has one of the darkest skies in Europe. One felt like one could just put out one's hands and touch the stars.

I had been awarded time on the JKT to make observations of some relatively nearby galaxies through different filters in the visible light part of the spectrum. The plan was to observe each of my target galaxies through U,B,V,R and I filters, which correspond to the near ultraviolet, the blue, green, red and very near infrared parts of the spectrum. By comparing the appearance of the same galaxy through these different filters, I hoped to gain a better understanding of how much the starlight was being affected by interstellar dust, which can very effectively hide stars towards the blue end of the visible spectrum, but its effects are less for starlight towards the red end of the visible spectrum. I did not know it at the time, but Kapteyn, after whom the telescope had been named, had played an important role in discovering the existence of interstellar dust in our Galaxy.

Jacobus Kapteyn was born in 1851 in the small Dutch town of Barneveld, about 40 km to the east of Utrecht. In 1868, he went to the University of Utrecht to study mathematics and physics. After completing his Thesis in 1875, he spent 3 years working at Leiden Observatory, before being appointed the first Professor of Astronomy and Mechanics at the University of Groningen in 1878.

At the time, the University of Groningen did not have an observatory, so Kapteyn collaborated with astronomers who did. Between 1896 and 1900 he worked with David Gill of the Royal Observatory at the Cape of Good Hope in South Africa, who had obtained photographic plates of the stars of the Southern Hemisphere. Kapteyn and Gill measured the positions and magnitudes of 454,875 stars in this painstaking work. As part of the study, in 1897 Kapteyn discovered what was, at the time, the star with the largest proper motion, which we now call "Kapteyn's star" [1]. It remained the star with the largest proper motion until the discovery in 1916 of Barnard's star. Just to remind you, a star's proper motion is an actual change in its celestial position over time, rather than the annual back and forth motion (stellar parallax) due to our orbit about the Sun.

2.1 Kapteyn's Universe

In 1904, continuing his study of the proper motion of stars, Kapteyn noticed that their motions were not random as was thought at the time [2]. Instead, he found that stars fell into two categories, with their proper motions in nearly opposite directions. Although he didn't realise it at the time, what he had found was the first evidence of the orbit of stars (including our Sun) around the centre of the Galaxy. It was his countryman Jan Oort who conclusively showed this in the late 1920s [3]. The proper motion of stars that we observe is precisely because stars are orbiting the centre of the Milky Way, and their speed of orbit, at least in our part of the Galaxy, falls off with distance from the centre of the Milky Way. Therefore, stars inside of our orbit will be travelling faster than us and appear to have proper motions in one direction, and stars further from the centre than us will be moving slower and appear to have proper motions in the opposite direction.

Two years later, in 1906, Kapteyn launched his most ambitious project, a major study of the distribution of stars in the Milky Way. He divided the Milky Way into 206 zones, and within each zone he measured the apparent magnitude, spectral type, radial velocity and proper motion of as many stars as he could. To undertake this massive task he used photographic plates from over 40 different observatories; it was the first ever coordinated statistical analysis in astronomy. To determine the actual distance of the stars, Kapteyn used a combination of the parallax and proper motions of stars. The completed work was finally published in 1922, entitled *"First attempt at a theory of the arrangement of motion of the sidereal system"* [4]. In this work, Kapteyn concluded that our Sun was in a galaxy shaped like a lens, thicker in the centre than at the edges, and with the Sun lying near the centre. He calculated the Galaxy to have a diameter of about 50,000 light years (17 kiloparsecs [kpc]), and a thickness at the centre of about 10,000 light years (3 kpc)—see Fig. 2.1. He believed the Sun lay about 2,000 light years (600 kpc) from the centre, slightly above the mid-plane.

As we shall see later, Kapteyn's model was wrong, but this was not due to any sloppiness on his part. Rather, it was due to his being unaware of the effects of interstellar dust in our Milky Way, the same dust I was trying to study using the telescope named after him. Interstellar dust scatters and absorbs starlight, and this leads to an effect called *"interstellar extinction"*, which effectively reduces the brightness of stars. It also reddens them (or, strictly speaking, "de-blues" them, no red light is added, but blue light is removed). Blue light is preferentially absorbed and scattered by interstellar dust, red light is less affected. If there is enough dust, a star can be rendered invisible to us. A well known example of interstellar extinction is the famous *Horsehead Nebula* in the Orion constellation, the nebula is hiding stars

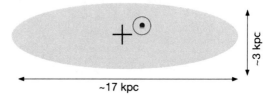

Fig. 2.1 A schematic of "Kapteyn's Universe", which placed the Sun near the centre of our Galaxy

Fig. 2.2 The Horsehead nebula in the Orion constellation (image credit—NASA)

which lie behind it, so the dark appearance of the "horsehead" is not the absence of stars, but rather their obscuration by a big cloud of foreground gas and dust (see Fig. 2.2).

As a consequence of the interstellar extinction due to dust, most of the stars in the plane of the disk of our Galaxy are not visible to us when we observe at wavelengths which lie in the visible part of the spectrum. So effective is dust at hiding stars that we can only see roughly the same distance in every direction, a distance of about 4 kpc (13,000 light years). Kapteyn correctly argued that seeing the same number of stars in all directions would suggest that we lie at the centre of the disk of stars, but he was misled because he did not include the effects of dust. This is strange, because in a paper in 1904 he had speculated on the absorption of light by dust [5], long before the effect was conclusively shown to be present by Trumpler in 1930 [6]. In fact, when we look in the near infrared, a wavelength long enough to not be affected by dust, we indeed find that the number of stars does vary as we look in different directions (see Fig. 2.3, an infrared image taken in the early 1990s). We now know that we lie about 8 kpc (about 26,000 light years) from the centre of our Galaxy in the mid-plane of the disk. Our position in the Galaxy, however, was not really settled until the late 1920s.

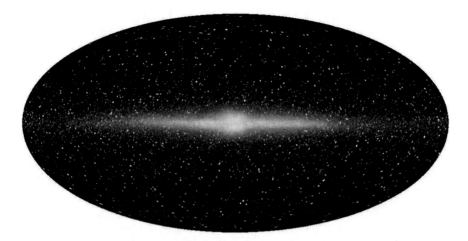

Fig. 2.3 An infrared picture looking towards the centre of our Milky Way galaxy shows it to be a flattened disk with a central bulge (image credit—NASA/COBE)

2.2 Galaxies Rushing Away

Vesto Slipher spent his entire career at Lowell Observatory, in Flagstaff, northern Arizona. The Observatory had been established in 1894 by Percival Lowell, the son of a wealthy Bostonian family. Lowell had graduated from Harvard in 1876, with a distinction in mathematics. After graduating, he ran a cotton mill for 6 years before spending time in the Far East of Asia. However, Lowell's passion was astronomy, and in the early 1880s he used his family's considerable wealth to establish the Lowell Observatory, on a mesa just to the west of the town of Flagstaff.

Although this part of Arizona does not have any mountains, it is at an altitude of over 2,500 m (7,000 ft). This, combined with dark skies and many cloudless nights, made Flagstaff an ideal location to choose. We tend to think of Arizona as a hot and humid place, but Flagstaff can get very cold—something I discovered when I visited there in March of 1993 to find snow on the ground.

One of Lowell's obsessions was Mars. He had read Camille Flammarion's *"La planète Mars"*, and was also familiar with the observations Italian astronomer Giovanni Schiaparelli made during the close opposition of Mars in 1877 (see for example [7]). During this opposition, Giovanni sketched features which he called *"canali"* in Italian, which means "channels" in English. But, the term got mis-translated as "canals", and Lowell became one of the main supporters of the idea that Mars was inhabited by intelligent beings who had built canals to move water from the poles to the arid equatorial regions.

The 24-in. refractor Lowell had installed at his Observatory was used by him to make further observations of Mars and its "canals". Lowell published three books on the subject of intelligent life on Mars, *"Mars"* (1895), *"Mars and its Canals"* (1906) and *"Mars As the Abode of Life"* (1908). However, in 1909 the theory of canals on

Mars was dealt a fatal blow by observations made with the new 60-in. reflecting telescope at Mount Wilson Observatory in Southern California. The images taken with the 60-in. were sufficiently detailed to show that the "canals" showed a number of irregular geological features, suggesting they were natural and not created by intelligent beings after all.

In an attempt to regain the Observatory's shattered reputation, Lowell started several new observing programmes in the 1910s making use of the 24-in. refractor. Lowell directed Slipher to head up one of these programmes—an attempt to measure the radial velocities of some "spiral nebulae". Slipher was relatively new to the Observatory, and had arrived after the whole "Mars debacle" had died down a little. Slipher had been born in 1875 in Mulberry, Indiana, and had obtained a Ph.D. from Indiana University in 1909. Upon completing his Ph.D. he went to work at Lowell Observatory, where he remained for the rest of his career.

When Slipher began his study of the radial velocities of spiral nebulae, little was known about their properties. Some astronomers argued that they were clouds of gas which were in the process of forming stars and planets, and lay firmly within our Galaxy. Other astronomers argued that they were "*island universes*", great stellar systems lying beyond our Milky Way whose stars were too distant to be seen as individual stars. The term "island universe" (what we would now call a "galaxy") was coined by the German philosopher Emmanuel Kant in 1755, and ever since a debate had raged as to whether these spiral nebulae were part of our Galaxy or beyond it. Galileo had shown back in the early 1600s that diffuse light could actually be individual stars. The band of diffuse light which stretches across the sky had been known since the dawn of civilisation as the Milky Way, but through a telescope Galileo was able to resolve this misty light into individual stars. Slipher would study the radial velocities of these spiral nebulae by measuring their spectra and trying to see how much the spectral lines had moved from their positions if the objects were not moving. This uses a phenomenon called the "*Doppler shift*".

The Doppler shift is something with which we are all familiar, it is what causes the sound of a siren to change pitch when an ambulance passes us. When the ambulance is approaching us, the sound waves coming from the siren have a higher frequency and so the pitch of the siren is higher, and when the ambulance is moving away from us the sound waves have a lower frequency and the pitch is lower. Light also travels as a wave, and so exhibits the same effect, but this time it is the colour of the light which is affected. When an object moves towards us the light is said to be "*blueshifted*", because the wavelength becomes shorter than if the object were at rest. If the object is moving away from us the light is said to be "*redshifted*", as the wavelength becomes longer than if the object were at rest.

This Doppler shift of stars' spectral lines allowed astronomers to determine whether stars were moving towards us or away from us, and what the speed of that motion was. It was found that there was no preferred direction for stars, some were moving towards us and some away from us. Astronomers also wanted to look at the spectra of "*nebulae*", the gaseous clouds which dotted the sky, but were hampered by their diffuse nature and their faintness.

2.2 Galaxies Rushing Away

In the 1880s and 1890s, they were just too faint for the telescopes and photographic plates then available to be able to record their spectra. But, by the 1910s with larger telescopes and more sensitive photographic plates being available compared to previously, Slipher and Lowell felt that taking spectra of these faint nebulae might be feasible, although to do so would require exposures of many hours, sometimes spanning several nights.

Slipher started taking his first spectrum in 1912, and chose the brightest spiral nebula in the sky, the *"Andromeda Nebula"*, also known as "Messier 31". He published his results in 1913 [8], and what he found was quite surprising. He found that the Andromeda Nebula was moving towards us (blueshifted), with a velocity of 300 km/s, much higher than that of any known star. The following year, Slipher published a paper which has been largely overlooked, in which he reported evidence that the Virgo Nebula (NGC 4594) was not only receding at about 1,000 km/s, but was also rotating [9]. This was the first ever evidence that spiral galaxies, as we now know them, rotate.

By 1915, Slipher had increased the number of spectra of spiral nebulae to 15 [10], with eleven clearly showing velocities away from us (redshifted), and two clearly showing a blueshift (Messier 31, also known as NGC 224; and Messier 32, also known as NGC 221). A third spiral nebula, Messier 33 (NGC 598), also showed a blueshift, and finally Messier 51a (NGC 5194—the "Whirlpool nebula") had a radial velocity that was too small to measure.

Of the galaxies that showed a redshift, most seemed to be to the North side of the band of the Milky Way, and Slipher wondered whether this was because the spiral nebulae were passing across the Galaxy in some way, moving in one direction in the North part of the Milky Way and the opposite direction in the South. As you can see from Fig. 2.4, several of the galaxies (NGCs 1068, 4565 and 4594) showed redshifts of at least 1,000 km!

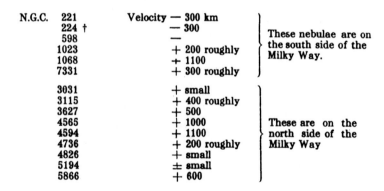

Fig. 2.4 A table from Slipher's 1915 paper showing the redshifts of 15 nebulae

Also in this 1915 paper, Slipher had an insight into the true nature of spiral nebulae. He noted that the spectra of nebulae which he had been gathering fell broadly into two classes, (1) "bright-line" spectra and (2) "dark-line" spectra (what we would now call "emission line" and "absorption line" spectra respectively). He stated [10]

> The so-called gaseous nebulae are of the first type, the spiral nebulae of the second type.

We now know that the "gaseous nebulae" he refers to are areas of ionised gas within our Galaxy, such as the Orion Nebula, which have bright emission lines because the gases are being excited by energetic photons from young stars. The "spiral nebulae", on the other hand, are galaxies outside of our own Galaxy, and their nebulosity is due to the stars from which they are comprised being too far away to see as individual stars, but each star is producing an absorption spectrum as the photons from the surfaces of the stars pass through the atmospheres of the stars. But, no one knew that at the time.

By 1917 Slipher had increased his sample still further, to 25 spiral nebulae [11]. As can be seen from Fig. 2.5, now 21 of the 25 spiral nebulae showed a redshift, with only four showing a blueshift. Obtaining these spectra required painstaking work. Exposures were often 20–40 h in length, requiring the photographic plates to be exposed over several nights, or if the weather was poor, several weeks.

Such observing could also only be done when there was no Moon in the sky, as the brightness of the Moon severely affects one's ability to observe such faint objects. As someone who started my astronomy career in the late 1980s, I have always observed sitting in a nice warm control room whilst a computer takes care of moving the telescope across the sky to follow objects for hour after hour. The hardest part is deciding what music to listen to in the comfort of my seat in front of

TABLE I.

RADIAL VELOCITIES OF TWENTY-FIVE SPIRAL NEBULÆ.

Nebula.	Vel.	Nebula.	Vel.
N.G.C. 221	− 300 km.	N.G.C. 4526	+ 580 km.
224	− 300	4565	+1100
598	− 260	4594	+1100
1023	+ 300	4649	+1090
1068	+1100	4736	+ 290
2683	+ 400	4826	+ 150
3031	− 30	5005	+ 900
3115	+ 600	5055	+ 450
3379	+ 780	5194	+ 270
3521	+ 730	5236	+ 500
3623	+ 800	5866	+ 650
3627	+ 650	7331	+ 500
4258	+ 500		

Fig. 2.5 A table from Slipher's 1917 paper, he had increased the number of nebulae to 25

the computer. Since 1996, when I started using the 3.5 m telescope at Apache Point in New Mexico remotely, I have been able to observe from the comfort of my study, often in my pyjamas!

In contrast, it is hard to imagine how astronomers like Slipher and others of his day must have suffered in making these long exposures. They had to guide the telescope by eye; whilst the photographic plate was being exposed through the main telescope they would sit looking through a smaller telescope mounted to the side of the main telescope, and adjust the position of the telescope minute by minute to compensate for any irregularities in the motion of the motors driving the telescope's basic tracking.

The average radial velocity of the twenty one receding nebulae Slipher included in this paper was found to be 570 km/s, which was about 30 times the typical radial velocity of stars! This high velocity, combined with the fact that most were moving away, was clearly telling us something, and Slipher struggled to interpret it.

In this 1917 paper [11] Slipher states

> Referring to the table of velocities again: the average velocity 570 km is about thirty times the average velocity of the stars. And it is so much greater than that known of any other class of celestial bodies as to set the spiral nebulae aside in a class to themselves. Their distribution over the sky likewise shows them to be unique - they shun the Milky Way and cluster about its poles.......... It has for a long time been suggested that the spiral nebulae are stellar systems seen at great distances. This is the so-called 'island universe' theory, which regards our stellar system and the Milky Way as a great spiral nebula which we see from within. This theory, it seems to me, gains favor in the present observations.

2.3 Herber Curtis, Harlow Shapley, and the Great Debate

The nature of spiral nebulae, and whether they lay outside of our Galaxy or not, was probably the most important problem in astronomy in the second decade of the twentieth century. The *American National Academy of Sciences* decided to organise a debate on the topic at the Smithsonian Institute in Washington, DC in April 1920. The two astronomers chosen for the debate could not have been more different, Herber Curtis and Harlow Shapley. The official title of the debate was "*The Scale of the Universe*", but it has come to be known as "*the great debate*". At the time, Curtis was 48 and worked at the Lick Observatory in Northern California, Shapley was 35 and worked at the Mount Wilson Observatory in Southern California.

Curtis studied at the University of Michigan and then at the University of Virginia, where he obtained a degree in astronomy. He had been working at Lick Observatory since 1902, continuing the work of surveying nebulae that James Keeler had begun in the 1890s using the 36-in. refractor. In 1912 Curtis was elected President of the Astronomical Society of the Pacific, and at the time of the Debate was one of the most important observational astronomers in the US; a polished speaker, confident, charismatic and well liked by his colleagues.

In contrast, Shapley was over a decade younger than Curtis. He had grown up poor in rural Missouri on a farm, and dropped out of school with only a fifth-grade education. After spending a few years working as a crime reporter for a local newspaper, he returned to complete his high school education, graduating top of his class. At the age of 22, Shapley decided to go to the local University of Missouri to study journalism. However, the opening of the School of Journalism had been postponed for a year, so he searched for another course to take. As he scanned through the course directory, he skipped past Archeology, which he later said he could not pronounce, and came to Astronomy.

This accidental astronomer had chosen well, he excelled at the subject. After graduating with high honours, Shapley went to Princeton for post graduate study under the direction of Henry Norris Russell, probably the most famous astrophysicist in the United States at the time, and co-discoverer of what we now call the "*Hertzsprung-Russell diagram*" for stars. At Princeton, Shapley worked on determining the distances to globular clusters [12, 13], nebulae that resolved into dense swarms of stars when viewed through telescopes (see for example, Fig. 2.6, which shows the globular cluster Messier 13). To obtain their distances, Shapley used the newly discovered period-luminosity relationship for Cepheid variable stars that I discussed in Sect. 1.11, and as part of this he did important work which helped in the understanding of Cepheids [14]. With this method of determining distances, Shapley was able to precisely measure the distances to globular clusters, and by virtue of this excellent work in 1914 he was hired by George Ellery Hale to a position at Mount Wilson Observatory, where he would soon have access to the

Fig. 2.6 Messier 13—an example of a globular cluster (image credit—NASA)

largest telescope in the World, the Hooker 100-in., to continue his globular cluster research.

By 1920, Shapley had established himself as a well respected astronomer, but he was still a nervous person, and felt overawed at the prospect of debating the charismatic Curtis in front of a live audience of colleagues and other scientists, which even included Einstein, at the Smithsonian. As fortune would have it, they ended up on the same train from California to Washington, DC, but Shapley did his best to avoid engaging in any conversation with Curtis during the 4,000 km trip to the East Coast. He spent the journey nervously rehearsing his talk, whilst Curtis chatted with fellow passengers during the several days the journey took.

The Great Debate focused on three topics

1. Whether spiral nebulae were within or outside of the Milky Way
2. The position of the Sun and Earth within the Milky Way
3. The size of the Milky Way

Clearly the first and third topic were, in many ways, related, as a very large Milky Way would lend support to the argument that spiral nebulae were part of it. The two astronomers held opposite views on each of these three questions. The Debate took the form of each man in turn giving his argument for all three topics before the other gave his, so didn't really take the form of a debate at all. The name came about over time, but the actual evening was two lectures from these two distinguished astronomers on each of the three topics.

Shapley began, and started by arguing that nebulae were within the Milky Way galaxy. He based his argument on two pieces of evidence. The first was where nebulae were found in the sky, which was generally above and below the disk of the Milky Way, but rarely within the disk. Shapley believed the nebulae to be clouds of gas in the process of forming stars and planets. As stars were found in the disk of the Milky Way, Shapley argued that they formed above and below the disk, and drifted into the disk once they had formed. If the nebulae were galaxies beyond the Milky Way as Curtis would later argue, Shapley said they should be found all over the sky as they were just as likely to lie beyond the disk of the Milky Way as they were above and below it.

The second piece of evidence Shapley used to support his view that nebulae lay within the Milky Way was based on a "*nova*" that appeared in the Andromeda Nebula in 1885 [15]. The term "nova" is, we now know, a misnomer. A nova is not a new star at all, but rather a star which briefly flares up and becomes brighter. The nova of 1885 in Andromeda was about one tenth as bright at the Nebula itself, and this would make sense if Andromeda were just a small collection of stars situated in the outer parts of our Galaxy, as Shapley believed. If, however, Andromeda were a galaxy in its own right, as Curtis and his supporters argued, it would presumably be made up of millions of stars, just like our Galaxy was. How could a single star be one tenth the brightness of millions of stars? It didn't make sense to Shapley, so the only rational conclusion he argued was that the Andromeda Nebula and the all the other spiral nebulae were part of our Galaxy.

Shapley used his observations of globular clusters, and the distances he had determined for them using Cepheid variables, to put forward his arguments on the size of the Galaxy and our position in it. Shapley argued that the distribution of globular clusters in the sky, where more were found in the direction of Sagittarius than in the opposite direction, suggested that they were distributed spherically around the plane of the Milky Way, with the Sun off to the side of the centre (see Fig. 2.7). Finally, Shapley used the Cepheid variables he had found in globular clusters to argue that the Milky Way was about 100 kpc in diameter.

After Shapley had presented his arguments it was Curtis' turn to present his. Curtis had determined a much smaller size for the Galaxy, and did not believe that the period-luminosity relationship for Cepheid variables applied to stars within the Galactic disk. Curtis instead based his arguments for the size of the Milky Way on star counts and distance estimates based on looking at the spectral types of stars and making assumptions about their intrinsic brightness. His model for the Galaxy was very similar to Kapteyn's, with the Sun near its centre and with a diameter of about 10 kpc.

Curtis believed that spiral nebulae were outside of the Milky Way, and represented "island universes" in their own right, similar in size to our Galaxy. He dismissed Shapley's argument that they were only distributed above and below the disk of the Milky Way, suggesting instead that nebulae were indeed scattered all

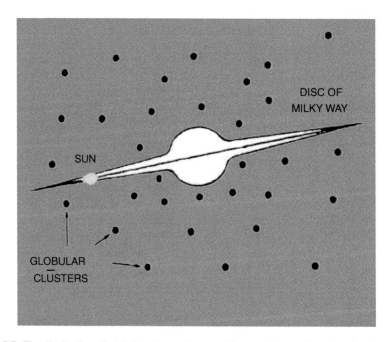

Fig. 2.7 The distribution of globular clusters in space. Because they are found predominantly in the direction of Sagittarius, Shapley argued that this meant that the Sun must be away from the Galaxy's centre

over the sky. The reason few nebulae were seen in the disk of the Milky Way, Curtis argued, was because the stars and gas and dust in the Milky Way were blocking our view of them. As for the nova of 1885, Curtis dismissed it as "abnormal", saying that other novae observed in the arms of spiral nebulae had all been much fainter than the one in Andromeda. In fact, Curtis claimed that the faintness of most novae showed that they were extremely distant and lay within nebulae that were too far away to be part of our Galaxy.

It is now interesting to look back on this debate with the hindsight of our current knowledge. We can conclude that the debate was essentially a draw. Curtis was correct about spiral nebulae lying outside of our Galaxy, but incorrect about the size of the Milky Way and the Sun's location in it. Shapley, on the other hand, was correct about our position in the Milky Way, but wrong about the nature of spiral nebulae. Both had got the size of the Galaxy wrong, with Curtis' 10 kpc diameter being too small, and Shapley's 100 kpc being too large.

The written summaries of the Great Debate were published by *The National Research Council* in 1921, entitled *"The Scale of the Universe"* [16], with Part One presenting Shapley's arguments and Part Two presenting Curtis' counter-arguments. Conclusive proof that spiral nebulae lay beyond the Milky Way would soon come along, using the Cepheid variables that Curtis was so reluctant to accept as a distance indicator.

2.4 Blinking Stars in the Andromeda Nebula

Sadly for Slipher's pioneering work on nebulae, his contribution has been largely overlooked. One can argue that he had observational evidence for spiral nebulae being outside of our Milky Way, evidence for their rotations, and evidence that the Universe was expanding. But this ground-breaking work has been eclipsed by a much more famous astronomer, Edwin Hubble. Hubble was described in Bill Bryson's book *"A Short History of Nearly Everything"* [17] as

> a large mass of ego

and it is maybe because of his large ego that Hubble is even known by many people outside the astronomical community, and Slipher is not known by many within it.

2.4.1 Edwin Hubble

Hubble was born in 1889 in rural Missouri. His family moved to Wheaton, Illinois the following year. He obtained a Bachelor of Science degree in 1910 from the University of Chicago, and then went to study at Queen's College Oxford for 3 years as a Rhodes Scholar. In 1913 Hubble's father died, so he returned to the US to look after his mother and siblings. In 1914 he returned to the University of Chicago to study astronomy as a graduate student, and was based at Yerkes Observatory.

Hubble used the 24-in. reflecting telescope which George Ritchey had designed and built and which was housed in the south dome of the Observatory, directly above the office I would occupy some 70 years later. With this telescope he studied the nebulae which were becoming such a hot topic of research in the 1910s. Although the 24-in. at Yerkes had the same aperture as the 24-in. refractor Slipher was using at the same time at Lowell Observatory, the Yerkes telescope was much better suited to observing diffuse extended objects like nebulae. This is because it had a focal ratio (the ratio of focal length to aperture) of f/4 compared to Lowell's f/16. The shorter focal length of the Yerkes reflector meant the nebulae appeared brighter on photographic plates than with the longer focal length refractor at Lowell.

By the third year of his Ph.D., Hubble had taken over one thousand photographic plates of nebulae, with each one needing an exposure of at least 2 h, and often longer. His Ph.D. thesis was submitted in 1917, with the title "*Photographic Investigations of Faint Nebulae*" [18]. Hubble finished his Ph.D. just as the United States was declaring war on Germany, and he volunteered for the US Army where he rose to the rank of Major, but he did not see any combat duty. After the war was over, he spent a year in Cambridge, where he renewed his studies of astronomy, before taking up a staff position in 1919 at Mount Wilson Observatory.

2.4.2 George Ellery Hale

Mount Wilson Observatory had been founded in 1904 by George Ellery Hale, possibly the most remarkable and one of the most important people in twentieth century astronomy, Hale founded the University of Chicago's Yerkes Observatory in the 1890s. Born into a wealthy Chicago family, Hale had studied physics at MIT and, as part of his final year project, had invented the spectroheliograph, an instrument that enables an observer to obtain a spectrum of different parts of the Sun's disk by moving a slit across the disk as the photographic film is moved in the opposite direction.

Hale's father built him his own private observatory in their family home's garden in Kenwood, a suburb of Chicago, which he named "Kenwood Astrophysical Observatory". Here he conducted research, mainly on the Sun, and at his father's expense equipped the Observatory with some of the best equipment of the day, including a 12-in. refracting telescope and instruments which were made by George Ritchey, a craftsman whom Hale senior employed to work full-time for his son. In 1891, when he was just twenty two he was approached by the President of the newly established University of Chicago, William Rainey Harper, to become director of a new observatory the University of Chicago wished to establish. Hale refused the offer, but in June of the following year Harper tried again, and this time Hale agreed to the offer.

Through his father's business connections Hale approached Charles Tyson Yerkes, and persuaded the transport magnate to fund building a new observatory which would have the largest telescope in the World—the great 40-in. refractor. Appealing to Yerkes' vanity in funding such a prestigious project, Yerkes' parting words to Hale as Hale left the meeting was "*I don't care what it costs, send me the bill!*" [19]. The 40-in. refractor saw first light in May of 1897, but by 1903 Hale had become tired of the frigid Wisconsin winters. He tried to persuade The University of Chicago to establish a remote observing station on the summit of Mount Wilson in balmy southern California. The University of Chicago refused his request, so in 1903 Hale quit as Director of Yerkes Observatory and struck out on his own to establish his new Observatory.

Hale was a master of fundraising, probably due to his spending time mixing with his father's businessmen friends in Chicago. In 1904 Hale received funding from the Carnegie Foundation which allowed him to build a 60-ft solar tower telescope, which was completed in 1908. In the same year the 60-in. reflector for which he had also obtained from Carnegie was completed. When it saw first light, the 60-in. became the largest telescope in the World, taking the tittle from the 40-in. at Yerkes. In the short period of 5 years Hale had established the premier observatory in the World, but he had only just got started. He next raised money to build a 150-ft solar tower which was completed in 1910, and then got money for what would become the 100-in. Hooker telescope, funded by local Los Angeles businessman John D. Hooker. The Hooker 100-in. finally went into operation in 1917, and was by far the largest telescope in the World with a light-gathering power nearly four times that of its nearest competitor, which was the 60-in. just down the hill! This wealth of observing facilities ensured that the best and brightest astronomers were being attracted to Mount Wilson to get jobs.

When Hubble arrived in 1919 he made no secret of his opinion that spiral nebulae lay outside of the Milky Way, which of course was in stark contrast to one of the Observatory's most celebrated astronomers, Harlow Shapley. There was always a tension between Shapley and Hubble, not only because they disagreed on the nature of spiral nebulae. Their contrasting personalities was also a source of conflict, whereas Shapley was shy and understated, Hubble was the exact opposite. It was probably a relief to both men when Shapley accepted the position of Director of Harvard College Observatory soon after the Great Debate of April 1920. They were both originally from rural Missouri, and Shapley had done nothing to modify his accent. Hubble, on the other hand, had returned form his extended stay as a Rhodes Scholar at Oxford with an affected English accent, a habit of wearing plus-fours and a tweed jacket, and of smoking a pipe. It would be no overstatement to say that Shapley couldn't stand him, finding him pretentious and pushy.

During his first few years at Mount Wilson, Hubble showed himself to be a dedicated and accomplished astronomer, and got more and more observing time due to the quality of the photographic plates he was obtaining. He continued to study the nebulae that had formed the basis of his Ph.D. thesis, initially on the 60-in. and then more and more on the 100-in. as he moved up the pecking order at the Observatory.

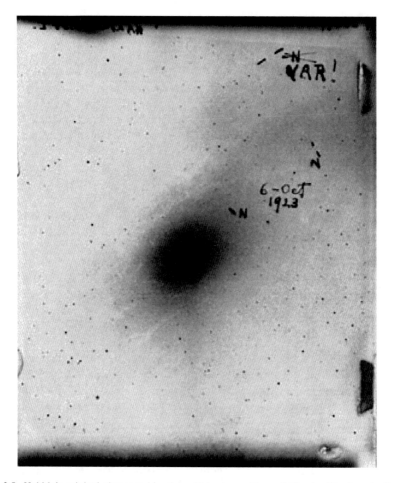

Fig. 2.8 Hubble's original photographic plate of his observations of Messier 31 where he finds a Cepheid variable. The "*VAR!*" is his designation for an object which he previously thought was a nova ("*N*") in fact being a Cepheid variable, allowing him to accurately determine for the very first time the distance to the Andromeda nebula

Access to the larger telescope put him in a rather unique position, able to study spiral nebulae to a degree of detail that no one else in the World was in a position to do.

In October 1923 Hubble was taking a photographic image of the Andromeda Nebula. It was the penultimate night of his observing run, and after developing the photographic plate the following day he noticed what appeared to be a nova in his image. He took another image the following night, and upon studying this one found a further two novae. He marked the three of them with an "*N*" (for *nova*)—see Fig. 2.8. He then went to the Observatory's photographic plate archive to check whether they were in fact novae, a nova would not show up on previous plates. Sure enough, he found that two of the "new stars" were indeed novae. However, the third

2.4 Blinking Stars in the Andromeda Nebula

one was even more interesting. Looking back through previous plates, he found that this third nova was sometimes present and sometimes not. It was not a nova at all, it was a Cepheid variable! This was a major discovery, as it would allow him to work out the distance to the Andromeda Nebula using Leavitt's period-luminosity relationship for Cepheids that I described in Sect. 1.11. He changed the designation of this object from "N" to *"VAR!"* on the photographic plate.

When Hubble did the calculations of the implied distance to this Cepheid using Leavitt's formula, the result was truly astonishing, and would change our understanding of the Universe for ever. The distance determined by Hubble for this Cepheid variable was 300,000 parsecs, more than three times the size of the Milky Way galaxy that Shapley had suggested, and thirty times the size preferred by Curtis! Whichever of the two might have been correct about the size of the Milky Way, Hubble's discovery showed that the Andromeda Nebula was far too far away to be considered a part of our Galaxy, and had to lie beyond it.

The result was so shocking that Hubble realised he needed to verify it before he went public. Over the next few months he obtained several more photographs of Andromeda, and this extra work paid dividends. He discovered a second, fainter, Cepheid, which confirmed his initial findings. With this second Cepheid supporting his calculated distance, he decided the evidence was strong enough to go public, but before publishing it he sent a private letter to Shapley. As Shapley had helped calibrate the period-luminosity relationship for Cepheids it was difficult for him to argue with Hubble's findings. He briefly tried, arguing that Cepheids with a period of over 20 days were different in nature to ones with a shorter period, but he soon realised his arguments were futile. Shapley had been proved wrong on the biggest question of the day—the nature of spiral nebulae and the scale of the Universe. When he received the private letter of Hubble's discovery Shapley is purported to have said

> Here is the letter that has destroyed my Universe

Hubble's discovery was announced to the wider astronomical community in the 1924 meeting of the American Astronomical Society in Washington, DC

With this one discovery, Hubble proved that this particular spiral nebula was, in fact, a great system of stars just like our Milky Way, just as Curtis and others had argued [20]. Of course, as there were dozens and dozens of fainter, smaller spiral nebulae it followed that they too were great collections of stars beyond our own Galaxy. Overnight our Universe became much vaster than most astronomers had previously imagined, and astronomers started referring to them as *"galaxies"* rather than "nebulae". Thus, for example, the *"Andromeda nebula"* was renamed the *"Andromeda galaxy"*. Ironically, the one astronomer who never adopted the name "galaxy" for these spiral nebulae was Hubble himself!

2.5 Hubble, Humason and an Expanding Universe

Hubble's discovery catapulted him to celebrity status. It was not long before he and his wife Grace were being invited to dine with many of Hollywood's "A-list" celebrities like Douglas Faribanks and Charlie Chaplin. Hubble revelled in this celebrity status, regaling anyone who would listen with what became increasingly exaggerated stories of his past. He exaggerated stories of his sporting prowess, of his upbringing, of his time in the Army (although he had never seen active service during the Great War, he was happy to tell people stories which often gave them the impression that he had taken on the Kaiser single-handedly in the trenches of Flanders!) But, through all of this, he did not forget what had made him famous in the first place, and continued his studies of spiral galaxies with relentless drive.

Hubble was aware of Slipher's work on the redshift of "nebulae", a finding which continued to confound astronomers. Over the next 6 years he and his observing assistant Milton Humason painstakingly obtained spectra of as many spiral galaxies as they could. Humason had worked his way into being involved in one of the most exciting astronomical research projects of the twentieth century from very humble beginnings. With barely any education, he had driven mules up to Mount Wilson with the raw materials which built the Observatory. From this menial job he showed that he was a hard worker, and so Hale gave him a slightly less menial job working at the Observatory itself. Slowly Humason impressed each and every member of staff at the Observatory with his hard work, his tireless attention to detail, and his affable manner. By the 1920s he was the main observing assistant on the biggest telescope in the World, and working alongside Hubble in studying far flung galaxies.

Even with the World's largest telescope at their disposal, obtaining spectra still required long exposures, often spanning several nights. Humason undertook much of this work, as it was boring and often cold. Hubble felt sufficiently comfortable with Humason's skills and diligence that he would often leave him in charge of the telescope for most of the night and would sneak off to get some sleep, as he wanted to be awake in the day to develop the completed photographic plates and analyse the images looking for Cepheids, and analysing the spectra. Obtaining spectra takes much more time that taking an image, as the light has to be passed through a narrow slit and then a dispersing medium spreads the light of different wavelengths out across the photographic plate. Whereas with a photograph all the light of a galaxy would fall on a small patch of the photographic plate, with a spectrum it might be spread out over an area tens of times larger, which dilutes the strength of the signal by the same amount. So, as a rule of thumb, if a direct image of a faint galaxy were to require a 2-h exposure, it would typically take about 20-h to obtain its spectrum!

Another complication which added to the time it took to obtain their spectra was the large mirror. The 100-in. mirror was made in the 1910s, at a time before low expansion glass was available. In fact, the glass below the silvered reflective surface just looks like bottled glass. I have used the telescope on four separate observing runs in late 1999 and early 2000, and so have used it for a total of about two dozen nights. On each occasion, we had to wait several hours after nightfall before we

2.5 Hubble, Humason and an Expanding Universe

could start out observations. The problem is that the mirror changes its shape as it cools, and so during the day time and a few hours into the night it is so out of shape that the images it provides are unusable. Then, as the Dome and the mirror cool down in the first couple of hours after sunset, the mirror slowly alters its shape to the right configuration to sharply focus the light that it was built to gather (Fig. 2.9).

By 1929 the pair had obtained the redshifts of 46 galaxies, and Hubble had estimated the distances of twenty of them. Most of the galaxies were too far away to be able to see Cepheid variables in them, so Hubble had to use less precise techniques, essentially "guesstimating" their distances. Some of these assumptions were things like assuming the brightest stars in each galaxy had the same intrinsic brightness, and even assuming that the total brightness of all similar looking galaxies were the same.

Despite these uncertainties in the distances, Hubble found an intriguing relationship between the much more precisely measured redshifts and the estimated distances. A plot showed that more distant galaxies appeared to be moving away more quickly, as indicated by their redshifts. The work was published in January 1929 with the title "*A Relation Between Distance and Radial Velocity Among Extra-Galactic Nebulae*" [21]. The original plot from this paper showing the relationship is shown in Fig. 2.10.

The most simple explanation of this relationship is that the Universe is expanding. To understand this, consider having a 30-cm ruler made of rubber or elastic. When it isn't stretched, the distance from the "0" mark to the "1" mark is, of course, 1 cm. The distance from the "0" to the "2" is 2 cm. If we stretch the ruler, so that the distance form the "0" to the "1" mark becomes 3 cm, the distance from the "0" to the "2" mark will now be 6 cm. Let us suppose that we take one second to do this stretching, the "1" mark will move from being 1 cm away to being 3 cm away, a difference of 2 cm, so it moves at 2 cm/s. The "2" mark will have moved from 2 cm to 6 cm, a difference of 4 cm, so it moves at 4 cm/s, twice the speed of the "1" mark. This is exactly what appears to be happening to galaxies based on Hubble and Humason's finding.

The speed of the expansion was determined by Hubble to have a value of about 500 km/s/Mpc, so a galaxy at 1 Mpc would move away at 500 km/s, a galaxy at 5 Mpc would move away at 2,500 km/s. This can be determined from the gradient (slope) of Fig. 2.11, and has been given the name "*the Hubble constant*". Despite his high opinion of himself, the "large mass of ego" was never one to interpret his observations too much. He felt his job was to make the observations and to present them, and allow others to derive theories based on those observations. Actually, this attitude was possibly a result *of* his ego, Hubble very much considered himself the World's greatest observational astronomer, he did not want to dilute his self-perceived preeminence by sullying his hands doing theoretical work!

At no point in this 1929 paper does Hubble mention that he and Humason had discovered that the Universe is expanding. But, of course, every one else realised that it was indeed what he had done. Even Einstein came to Mount Wilson to visit the great Hubble.

Fig. 2.9 The 100-in. Hooker Telescope at Mount Wilson. Even with the vast light-collecting power of the World's largest telescope, obtaining the spectra of faint galaxies often took several nights of observing

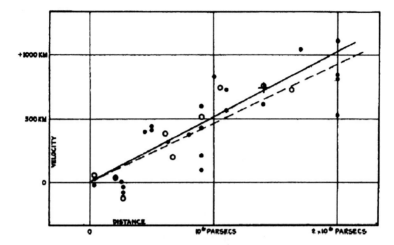

Fig. 2.10 Hubble's original plot of his "Velocity-Distance Relation among Extra-Galactic Nebulae" as published in his paper in January 1929 [21]

Fig. 2.11 In 1931 Einstein visited Mount Wilson Observatory. Here he is peering through the eyepiece of the 100 in., with Hubble hovering in the background. The Observatory Director, Walter Adams, is wearing the flat cap. Image credit—California Institute of Technology

References

1. Kappten, J.C.: Stern mit grösster bislang bekannter Eigenbewegung (A star with the greatest of previously known proper motions). Astronomische Nachrichten **145**, 159 (1897)
2. Kapteyn, J.C., & Desseter, W.: The proper motions of the Hyades, derived from plates prepared by Prof. Anders Donna. In: Publications of the Kapteyn Astronomical Laboratory Groningen, vol. 14, pp. 1–87. Hoitsema Brothers, Groningen (1904)
3. Oort, J.H.: Observational evidence confirming Lindblad's hypothesis of a rotation of the galactic system. Bull. Astron. Inst. Netherlands **3**, 275–282 (1927)

4. Kapteyn, J.C.: First attempt at a theory of the arrangement and motion of the sidereal system. Astrophys. J. **55**, 302–328 (1922)
5. Kapteyn, J.C.: Remarks on the determination of the number and mean parallax of stars of different magnitude and the absorption of light in space. Astrophys. J. **24**, 115–122 (1904)
6. Trumpler, R.J.: Preliminary results on the distances, dimensions and space distribution of open star clusters. Lick Obs. Bull. **14**(420), 154–188 (1930)
7. Washam, E.: Cosmic errors: Martians build canals! Smithsonian Mag. (2010). http://www.smithsonianmag.com/science-nature/lunar-bat-men-the-planet-vulcan-and-martian-canals-76074171/
8. Slipher, V.M.: The radial velocity of the Andromeda Nebula. Lowell Obs. Bull. **1**, 56–57 (1913)
9. Slipher, V.M.: The detection of nebular rotation. Lowell Obs. Bull. **2**, 66 (1914)
10. Slipher, V.M.: Spectrographic observations of nebulae. Pop. Astron. **23**, 21–24 (1915)
11. Slipher, V.M.: Nebulae. Proc. Am. Philos. Soc. **56**, 403–409 (1917)
12. Shapley, H.: Outline and summary of a study of magnitudes in the globular cluster messier 13. Publ. Astron. Soc. Pac. **28**, 171–176 (1916)
13. Shapley, H.: On the distribution of stars in globular clusters. Observatory **39**, 452–456 (1916)
14. Shapley, H.: Miscellaneous notes on variable stars. Astrophys. J. **41**, 291–306 (1915)
15. Hartwig, E.: Ueber den neuen Stern im grossen Andromeda-Nebel (About the new star in the great Andromeda nebula). Astronomische Nachrichten **112**, 355–358 (1885)
16. Shapley, H., Curtis, H.D.: The Scale of the Universe. Bulletin of the National Research Council, vol. 2, Part 3 #11, Washington (1921). Available online at e.g. https://archive.org/details/scaleofuniverse00shap
17. Bryson, B.: A Short History of Nearly Everything. Black Swan, London (2003)
18. Hubble, E.P.: Photographic Investigations of Faint Nebulae. The University of Chicago Press, Chicago (1917)
19. Osterbrock, D.E.: Yerkes Observatory, 1892–1950: The Birth, Near Death, and Resurrection of a Scientific Research Institution. University of Chicago Press, Chicago (1997)
20. Hubble, E.P.: Cepheids in spiral nebulae. Observatory **48**, 139–142 (1925)
21. Hubble, E.P.: A relation between distance and radial velocity among extra-galactic nebulae. Contrib. Mt. Wilson Observatory **3**, 23–28 (1929)

Chapter 3
The Cosmic Microwave Background

After Einstein published his general theory of relativity in 1916, it was not long before theoreticians used his equations to calculate how they applied to the Universe as a whole. Alexander Friedmann in 1923 and Georges Lemaître in 1927 independently found that one possible solution is that the Universe could be expanding, Lemaître even calculated an expansion rate based on Slipher's data. In 1931 Lemaître went on to propose the idea of a "primordial atom". In 1929 Hubble "discovered" the expansion of the Universe. Lemaître is very much the father of the "big bang theory", but Sir Fred Hoyle and others argued in favour of a competing theory, the "steady state theory". The first prediction of a cosmic microwave background was made by Alpher and Herman in 1948, after Alpher and Gamow had been working on the origin of the chemical elements. But, despite their prediction, it was decided by the radio astronomy community that it would be too hard to detect. It was discovered in 1965 completely by accident by Penzias and Wilson. This discovery led to the two being awarded the Nobel Prize in 1978, and pretty much put paid to the steady state theory.

3.1 The Primordial Atom

Hubble's discovery in 1929 of the expansion of the Universe led to one of Einstein's more memorable quotes. In 1916 Einstein published his General Theory of Relativity, his theory of gravity [1]. In 1917 he found that his equations led to an unstable Universe [2]. At the time, astronomers thought the Universe was static (neither expanding nor contracting), so Einstein introduced a "fudge-factor" to make the Universe stable and static—something which has come to be known as the "cosmological constant". When Hubble found that the Universe was, in fact, expanding, Einstein referred to his cosmological constant as "the biggest blunder of my life".

Although Einstein had failed to predict the expansion of the Universe, others had used his equations to make just such a prediction. The first was Dutch mathematician, physicist and astronomer Willem de Sitter. In 1917 de Sitter came up with a solution to Einstein's equations of General Relativity in which a Universe with no matter in it would expand [3]. However, everyone knew the Universe had matter in it, so this work was not treated very seriously.

A few years later, in 1922, Russian cosmologist and mathematician Alexander Friedmann (Fig. 3.1) derived solutions to Einstein's equations which predicted a dynamic Universe, but this time with matter in it [4]. He published this work in the German journal *Zeitschrift fṛ Physik* with the title *Über die Krümmung des Raumes* ("On the curvature of space"), and he was the first person to derive such solutions to Einstein's equations which predicted an expanding (or contracting) Universe which contained matter. *Zeitschrift für Physik* was a scientific journal founded by Albert Einstein and others in 1920 for original research articles, and so it is not true that Friedmann published his work in an obscure journal as some people have argued.

Friedmann was born in Saint Petersburg in 1888, his father was a composer and ballet dancer and his mother a pianist. He attended Saint Petersburg University, graduating in 1910, and served as an aviator and later aviation instructor for the Imperial Russian Army in the first World war. In 1918 he was offered a professorship

Fig. 3.1 In 1923 Alexander Friedmann developed a theoretical model of the Universe which predicted an expanding (or contracting) Universe with matter in it

at Perm State University, which is some 1,500 km (950 miles) virtually due east of Moscow in the Urals.

In 1923 Friedmann published a book in Russian of his ideas, the English translation of the title is "*The World as Space and Time*" [5]. It is one of the first attempts to popularise the ideas of General Relativity to a wider audience. The English translation of this book has only become available in March 2014. In this book, Friedmann states on page 79 of the English translation

> The non-stationary type of Universe presents a great variety of cases: for this type there may exist cases when the radius of the curvature of the world, starting from some magnitude, constantly increases with time; there may further exist cases when the radius of curvature changes periodically: the Universe contracts into a point (into nothingness), then again, increases its radius from a point to a given magnitude, further again reduces the radius of its curvature, turns into a point and so on. This unwittingly brings to mind the saga of the Hindu mythology about the periods of life; there also appears a possibility to speak about 'the creation of the world from nothing,' but all of that should be viewed as curious facts which cannot be solidly confirmed by the insufficient astronomical material.

Friedmann published a second landmark cosmology paper in 1924 [6], publishing in the same German journal *Zeitschrift für Physik* . But, sadly for the world of cosmology, Friedmann died a premature death in 1925 from typhoid fever which he had contracted during a holiday in Crimea.

A few years later, in 1927, a Belgian Catholic priest and astrophysicist Georges Lemaître (Fig. 3.2) independently came up with the same ideas of an expanding Universe with matter in it. Lemaître had obtained his degree in mathematics in 1920, and in 1923 was ordained as a priest. However, in the same year he went to work at Cambridge University where he came under the influence of Sir Arthur Eddington, the premier English astrophysicist of the day. Eddington introduced Lemaître to Einstein's theory of General Relativity and to the subject of cosmology. After a year in Cambridge, Lemaître went to work with Shapley at Harvard College Observatory, and in 1925 returned to his native Belgium to take up a part-time lecturing position at the Université catholique de Louvain.

In 1927 he published a paper in French in a Belgian scientific journal, the *Annales de la Société Scientifique de Bruxelles* outlining his ideas of an expanding Universe [7]. Aware of Slipher's results on the redshifts of nebulae, Lemaître proposed that these redshifts were signs of an expanding Universe. Not only did he propose an expanding Universe on theoretical grounds, but using the published data on the redshifts and distances of forty two galaxies, he calculated the rate of expansion, what we would now call the "Hubble constant". The rate he found was 625 km/s.

However, although French was a language widely read by scientists at the time, the journal in which Lemaître published his work was not, and so the work went largely unnoticed. In 1931, Lemaître's 1927 paper was translated and published in *The Monthly Notices of the Royal Astronomical Society* [8], one of the most widely read astronomical journals. To accompany the translation, Lemaître wrote an update on his work on the problem of an expanding Universe [9], and this was read out by Sir Eddington at the Royal Astronomical Society meeting in Burlington House, London.

Fig. 3.2 Georges Lemaître independently developed a model of an expanding Universe in 1927. In this paper he also calculated the expansion rate of the Universe based on Slipher's observations, and in 1931 predicted that the Universe began in a *"primordial atom"*

Although Eddington was one of the most well-respected cosmologists in the World, Einstein was not convinced with Lemaître's theory, and so it gained little acceptance. But, all that changed in 1931 when Lemaître published a letter in the prestigious scientific journal *Nature*, outlining his expanding Universe theory, and suggesting that the Universe had begun as a *"primordial atom"*. This time, newspapers around the World picked up on the article and he was hailed as the *"leader of the new cosmological physics"*. The New York Times ran the front page headline

> Lemaître suggests one, single, great atom, embracing all energy, started the Universe

The final seal of approval came when Einstein himself became convinced of the theory. In 1932, collaborating with de Sitter, the two of them found a solution to Friedman's equations where space was flat but expanding [10, 11]. This so-called "Einstein-de Sitter Universe" was one which would expand forever but would take an infinite amount of time to do so. With Einstein on board, the idea of an expanding Universe could move into the main stream of cosmological thinking.

3.2 Fred Hoyle and the Steady State Theory

Although Einstein was finally won over to embracing Lemaître's idea of a primordial atom, it was not the only cosmological theory around. Its main rival was a theory which became known as *"the steady state theory"*, and its main champion was Sir Fred Hoyle. Hoyle was one of the most influential cosmologists of the twentieth century. Born in 1915 and raised in Yorkshire in Northern England, Hoyle's father was a wool trader, and he grew up in financially insecure circumstances. His mother had attended the Royal College of Music in London, and it was she who taught Hoyle his multiplication tables at a young age. Hoyle did not like school, he found it difficult to conform and preferred to learn on his own. But, of course, the law required him to attend school, but there he would frequently argue with his teachers.

At the age of 13, Hoyle's father bought him a telescope, and it was in using this to discover the night-time sky that Hoyle found his true passion. He started reading books on astronomy and cosmology, including those by Eddington, who had become one of the most widely read popularisers of astronomy in Britain in the 1920s and 1930s. In his writings, Eddington had expressed his discomfort with Lemaître's "primordial atom" idea. As he had stated in a paper in *Nature* in 1931 [12]

> Philosophically, the notion of a beginning of the present order of Nature is repugnant to me. I should like to find a genuine loophole.

Eddington was to repeat this reservation in his popular writings. It is clear that this had a long-lasting affect on Hoyle, because in adulthood he became the main proponent of a competing cosmological theory, the steady state theory.

After finishing grammar school, Hoyle went to Emmanuel College Cambridge to study mathematics. In 1940, he left Cambridge to help in the war effort, working on radar systems in Portsmouth on the south coast of England. It was here that Hoyle was to meet fellow Cambridge graduates Hermann Bondi and Thomas Gold, and the three of them formed a close friendship. Bondi had studied mathematics at Trinity College Cambridge, and Gold had studied Natural Sciences, also at Trinity. One of their main topics of conversation was cosmology, one of Hoyle's passions.

After the war, Hoyle returned to Cambridge where he became a lecturer at Saint John's College, and he embarked on what was to become a very illustrious career in theoretical astrophysics research. In 1946, almost immediately upon his return to Cambridge, Hoyle started thinking about cosmology and the origin of the chemical elements, As far as cosmology was concerned, he was convinced that the Universe had not begun in a single event as Lemaître had argued, but rather had always existed and would always exist. The origin of the elements, he was convinced, was inside of stars.

In 1948 Bondi and Gold published a paper entitled *"The Steady-State Theory of the Expanding Universe"* [13]. Their argument in this paper was very much a philosophical one, saying that in order for the physical laws we find here on Earth to be true throughout the Universe, then the Universe had to have always been the same and could not be evolving and changing. Hoyle published a more mathematical paper entitled *"A New Model for the Expanding Universe"* in the same journal a few weeks later [14], in which he re-wrote Einstein's equations of General Relativity to include the idea of the continuous creation of matter, and showed that this did away with the need for the cosmological constant which Einstein had introduced in order to give a stable, stationary solution to his equations.

As we will see in Sect. 3.4, 1948 was also the year that George Gamow, Ralph Alpher and Robert Herman were publishing their papers championing the primordial atom theory. The battle lines would be drawn between the two competing theories for the next 20 years. Notice that so far I have referred to the Lemaître theory as the *"primordial atom"* theory, not the *"big bang theory"* as we call it today. That is because this term was not coined until March 1949, and it was Hoyle himself who coined it! In a BBC radio interview, he referred to the competing theory as "the big bang theory", it was probably meant to be a derogatory term but the name stuck. [15]

3.3 Modern Alchemy

It is said that the father of modern physics, Sir Isaac Newton, spent just as much time in the pursuit of alchemy as he did on his physics [16]. Alchemy is the ancient belief that worthless metals (and other materials) could be changed into gold and other precious substances. In fact, the word "chemistry" comes from the word "alchemy", with the *al* being from the Arabic word for "the". With the development of chemistry as a scientific discipline in the early 1800s, the idea of being able to change one element into another was discredited.

However, that was to change in 1896 when Henri Becquerel accidentally discovered radioactivity [17]. As scientists worked to understand this new process, it was realised by Ernest Rutherford in the early part of the 1900s that radioactive elements would change into a different element if they emitted an alpha or a beta particle, two of the three types of radioactivity (the third type, gamma decay, will leave an element unchanged unless it is accompanied by alpha or beta decay). An alpha particle was found to be the same as the nucleus of a helium atom, and so when e.g. thorium decays through emitting an alpha particle it changes into the element radium.

What has actually happened is that the thorium nucleus has lost two protons and two neutrons (which is a helium nucleus). As each element has a unique number of protons, a different number of protons means a different element. By losing two protons, the thorium becomes radium. Similarly, carbon-14, a radioactive form of carbon which is used in radioactive carbon dating of once living material, decays via beta decay into nitrogen. What happens in beta decay is that a neutron changes into a proton, and a high speed electron is spat out of the nucleus. By gaining an extra proton, the carbon becomes nitrogen. This is natural alchemy at work, but not really of the form envisaged by the adherents of ancient alchemy.

Rutherford was not just the person who untangled the details of radioactivity, he also discovered the atomic nucleus in a series of experiments in the period 1909–1912. After his discovery of the nucleus he went on to identify the proton as the positively charged particle in the nucleus, and in 1917 Rutherford was the first person to artificially change one element into another. Although the work would not be published for another 2 years [18], in a letter to Niels Bohr in 1917 [19] he stated

> I am trying to break up the atoms by this method....

"This method" was bombarding normal, stable nitrogen with alpha particles, and changed the nitrogen into oxygen when one of the two protons and both neutrons in the alpha particle stuck to the nitrogen nucleus. He thus showed that, through the bombardment of nuclei by other nuclei, elements could be built up. For the transmutation to work the helium nuclei had to hit the nitrogen nuclei at high speed, because the particles had to overcome their electrical repulsion as nuclei are always positively charged due to their protons. This is the basic idea of nucleosynthesis.

3.4 The Origin of the Chemical Elements

In the same year of 1917, the American chemist William Draper Harkins had noticed the curious fact that, with the exception of hydrogen, elements that had even atomic numbers (2, 4, 6, 8 etc.) were far more abundant that elements with odd atomic numbers (3, 5, 7, 9 etc.) [20,21]. Was this a clue to how elements had formed? With the nuclear transmutation work of Rutherford newly published, Harkins suggested that the heavier elements had been formed by the fusing (merging) of lighter elements, and that even numbered elements were more numerous because either they formed more readily or they were more stable.

Supporting evidence was found for this idea when, in 1925, English-American astrophysicist Cecilia Payne-Gaposchkin proposed in her Ph.D. thesis that the Sun was composed of mainly hydrogen [22]. Payne-Gaposchkin had studied Natural Sciences at Cambridge University, but when she completed her studies in 1923 she was not awarded a degree as Cambridge did not award degrees to women until 1948. She got a fellowship to go and study at Harvard, and under the guidance of Shapley she wrote her Ph.D. thesis. Although astrophysicists suspected that the Sun got its energy from nuclear power, it was initially assumed that it was from nuclear fission, where heavy elements spontaneously split into lighter ones. If the Sun were mainly hydrogen, fission wouldn't make sense. So the idea that was developed in the 1920s by Eddington and others was that the Sun was generating its energy by fusing hydrogen into helium. The details of this theory were worked out in the early 1930s by German-American theoretician Hans Bethe, e.g. [23, 24].

Could fusion also explain the origin of the other elements beyond helium? Maybe hydrogen and helium nuclei fused to form beryllium, and three helium nuclei fused to form carbon, etc. In his 1927 book *"Stars and Atoms"* [25] Eddington wrote

> The particles constituting helium's nucleus must have been assembled at some time and place; and why not in the stars?

But, by 1938, German physicist Carl Friedrich von Weizsäcker concluded that the Sun and stars were not hot enough to have fused these heavier elements [26]. He argued that the electrical repulsion between nuclei with so many protons was just too great. But, he suggested instead that such conditions may have existed in a primordial fireball.

This suggestion caught the imagination of Russian physicist George Gamow. Gamow was a polymath, who made contributions to physics, cosmology, and even biology. In the 1950s he was the first person to suggest how the various amino acids were built up from the four different bases (adenine, cytosine, thymine and guanine) that had been discovered in DNA by James Watson and Francis Crick [27]. Gamow came up with many ideas in his lifetime, some inspired and some crazy.

Born in 1904 in Odessa, Russia (now in the Ukraine), he studied initially at Novorossiya University in his home town, and then under Friedmann at the University of Leningrad. After graduating in 1929, he went to work on quantum theory at Göttingen in Germany, then at the Theoretical Physics Institute at the University of Copenhagen with Niels Bohr. His time in Copenhagen also included a brief spell at the Cavendish Laboratory in Cambridge, working under Rutherford. He returned to Russia to take up a position at the Physical Department of the Radium Institute in Leningrad.

However, Gamow started to become increasingly frustrated with what he felt were the restrictions put on scientists working in the Soviet Union. In 1931 he was denied permission to attend a conference in Italy, and resolved to leave Russia with his wife. In 1933 he was granted permission to attend the 7th Solvay Conference in Brussels, and obtained permission for his wife to travel with him, under the excuse that he needed her to act as his secretary.

With the help of Marie Curie, Gamow was able to extend his stay in Europe, and worked at the Curie Institute, the University of London and even the University of Michigan in the USA. Gamow never returned to Russia, and in 1934 he and his wife defected to the USA. He took a position professorship at George Washington University (GWU) in Washington D.C., and recruited Hungarian physicist Edward Teller from London to join him. Together they worked on some unresolved details of radioactive beta decay. Later, Teller would say of Gamow that

> ninety percent of Gamow's theories were wrong, and it was easy to recognise that they were wrong. But, he didn't mind. He was one of those people who had no particular pride in any of his inventions. He would throw out his latest idea and then treat it as a joke.

One of Gamow's Ph.D. students at GWU in the 1940s was Ralph Alpher, whom I had the privilege of meeting near the end of his career in 1995. Alpher was the son of Eastern European immigrants, and excelled at high school, particularly in maths. He graduated from high school at 15, but with his mother dying when he was only 17 he had to find himself paid work to help support the family. He took a course in shorthand, and got a job at the Carnegie Institute in Washington, becoming secretary to the Director. In 1940 he enrolled for evening classes at GWU, whilst in the day he was recruited into the National Ordnance Laboratory as part of the war effort.

After completing his Bachelors degree at GWU in 1943, Alpher requested and got Gamow to be his graduate studies advisor. He worked on his Masters thesis, which he was awarded in 1945, and at the end of the war transferred to working at the newly created Applied Physics Laboratory at Johns Hopkins University (JHUAPL), where Gamow was employed as a part-time consultant in addition to his position as a professor at GWU. Gamow suggested to Alpher that he study the origin of the elements for his Ph.D. thesis, which Alpher did on top of his full-time day job at the JHUAPL.

Alpher had done some work for the US Navy during the second World war, and after the war became acquainted with the new field of nuclear physics research which was going on at the Argonne National Laboratory just outside of Chicago and at the Brookhaven National Laboratory on Long Island. The data gathered from the experiments being done at these two particle accelerators proved vital in the work that Alpher and Gamow were working on—the building up of the elements during the first few moments of the Universe when it was hot enough and dense enough for "*neutron-capture*" to take place.

The basic idea Alpher and Gamow had was that hydrogen and helium were formed within the first 17 s after the beginning of the Universe, and that the other elements were created during an era of "*nucleosynthesis*" which lasted about 5 m. During this time, free neutrons would capture free protons in the "*primordial soup*", and in doing so build up heavier and heavier elements by changing via beta decay into protons. Gamow submitted the paper summarising Alpher's Ph.D. thesis work to the prestigious journal *Physical Review*, but ever the joker he mischievously added the name of Hans Bethe, so that the author list would read "Alpher, Bethe, Gamow" (alpha, beta, gamma) [28].

We now know that Alpher and Gamow were partly correct and partly incorrect. Hydrogen and helium are indeed created during a period of nucleosynthesis in the early Universe, but the heavier elements are not. They are built up within stars, something which Sir Fred Hoyle and others would show in the 1950s (e.g. [29]).

3.5 The Hot Surfaces of Stars

When we look into the night-time sky using our eyes, we of course see the stars, and the Moon and planets if they are visible. Why do the stars appear like points of light? Why are they not black like the rest of the night-time sky? The reason is that stars have surfaces which are at thousands of degrees Kelvin, and this temperature is hot enough to give off radiation that we can see with our eyes, what we call "visible light". But, in fact, you and I are also giving off radiation, it is just that our eyes are not able to see it.

A hot opaque solid, liquid or gas gives off radiation in a very particular way—something physicists call *"blackbody radiation"*. The name is a little misleading because a blackbody is not in black, the name comes from the fact that a perfectly black body is an ideal absorber and emitter of radiation. Stars radiate as blackbodies, and although the centres of stars are at millions of degrees, the surfaces are typically at thousands of degrees, and it is the radiation from the surfaces of stars that we can see with our eyes. Cooler objects, like our bodies, also radiate as blackbodies, but because the temperatures are lower (a few hundred Kelvin instead of a few thousand), the radiation does not come out in a part of the spectrum which we can see with our eyes, but rather in the *infrared* part of the spectrum. The infrared was, in fact, accidentally discovered by William Herschel in 1800 [30], and it is the radiation to which thermal imaging cameras are sensitive.

If one looks at the intensity of the radiation of a blackbody at different wavelengths (what scientists call a spectrum), it has a very particular shape. The shape of a blackbody's spectrum is the same no matter what the material of the blackbody, and no matter what the temperature. The blackbody can be a star of a heated cannon ball, the shape of the spectrum will look the same. The only thing that varies when studying the spectra of blackbodies is the wavelength (or frequency) at which the blackbody curve peaks, and the total area under the curve. These two things are related to the temperature, but the shape of the curves still look the same, as Fig. 3.3.

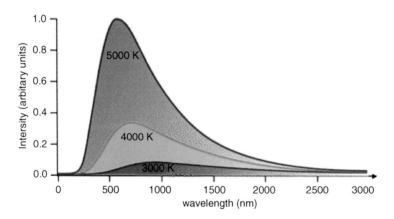

Fig. 3.3 The blackbody spectra of three stars at different temperatures. The *red curve* is for a star with a surface temperature of 3,000 K, the *green* for a star at 4,000 K and the *blue* for one at 5,000 K. The position of the peak is dependent on the temperature T, and the area under the curve is dependent on T^4

The relationship between the temperature of the blackbody and the wavelength at which the curve peaks was found by German physicist Wilhelm Wien in 1893 [31]. It is now known as *"Wien's displacement law"*, and it states that the wavelength (in metres) is just the number 0.002898 divided by the temperature T (in Kelvin). So, hotter objects will have their blackbody spectrum peaking at shorter wavelengths, and cooler objects at longer wavelengths (again, see Fig. 3.3). Objects at tens of thousands of degrees will radiate most strongly in the ultraviolet part of the spectrum, objects at millions of degrees will radiate most strongly in the X-ray part of the spectrum. Objects cooler than a few thousand Kelvin will radiate most strongly in the infrared, and as we go to lower and lower temperatures, the peak moves more and more towards the microwave and radio parts of the spectrum.

In 1934 American mathematical physicist Richard Tolman published a book entitled *"Relativity, Thermodynamics, and Cosmology"* [32] in which he showed that blackbody radiation in an expanding Universe would cool due to the stretching of space, but its spectrum would retain its characteristic blackbody shape.

3.6 Cooking Helium in the Early Universe

In the ht big bang model the temperatures in the early Universe would have been so hot that matter (protons, electrons and neutrons) would have interacting constantly with the radiation. It is this constant interaction between matter and radiation that, in fact, produces a blackbody spectrum. We say that the object is in *"thermal equilibrium"*. From Wien's law, the wavelength of the blackbody radiation's maximum intensity is determined by the temperature, and as the Universe expanded the temperature fell.

As Alpher, Herman and Gamow would argue [33], the early Universe is just a sea of protons, electrons and neutrons. Neutrons are unstable, if they are not attached to protons in an atomic nucleus they decay into a proton and an electron in about 14 min on average. If a proton could capture a neutron before it decayed it could form deuterium, a stable isotope of hydrogen which has a proton and a neutron in its nucleus. A deuterium nucleus could then capture a further neutron and form tritium, which is an unstable isotope of hydrogen but with two neutrons in its nucleus. Tritium could then capture another proton to form helium-4, the most common isotope of helium.

We now know that this is the way that helium is built up in the early Universe, but it is a race against time before the Universe has cooled too much for these reactions to take place. By the time the Universe is about 3 min old, it has cooled too much for the reactions to continue. The 25 % of helium which we see in the Universe was nearly all created in this first 3 min (Fig. 3.4).

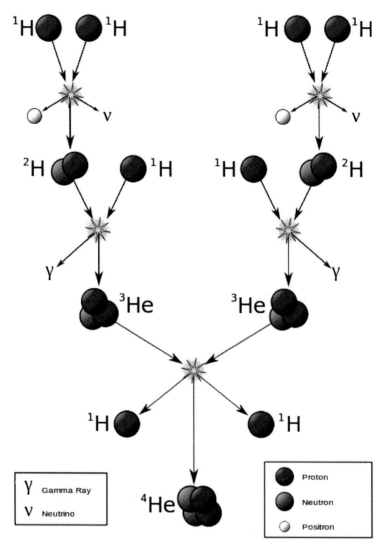

Fig. 3.4 Hydrogen is converted to Helium in the early Universe via a process known as the *"proton-proton chain"*

3.7 Cosmological Model Predictions

A scientific theory is only of any use if it not only explains observed phenomena, but also makes predictions against which it can be tested. Although both the steady state theory and the big bang theory argued that they could explain what was already known about the Universe, they also made predictions about things which were not known in the late 1940s and early 1950s. There were four predictions which were distinctly different for the two theories, and these were

1. The age of the Universe
2. The origin of the elements
3. The distribution of matter in both space and time
4. The temperature of the Universe

3.7.1 The Age of the Universe

On this question the steady state theory predicted that the Universe had always existed. This view of course harked back to the ideas of the ancient Greek philosophers, and in particular Aristotle. The big bang theory, on the other hand, predicted that the Universe had a beginning, and the simplest way to calculate the age of the Universe is to assume that its expansion has always been at the same rate, the so-called Hubble constant H_0.

If one assumes this, then the age of the Universe comes out to be just one divided by the Hubble constant, $1/H_0$. Unfortunately for supporters of the big bang theory, using the value of the Hubble constant which Hubble himself had determined in the late 1920s and early 1930s, a value of 500 km per second per Mega parsec (to state it in the usual units), gave an age to the Universe of only two billion years. This presented a problem as it was considerably younger than the age of the Earth that geologists had determined, which was over four billion years old. So, as things stood in the late 1940s and early 1950s, the big bang theory was not compatible with the observations.

3.7.2 The Origin of the Elements

Because Hoyle rejected the idea of a hot, early Universe, he was convinced that the elements beyond hydrogen had been made in the hot interior of stars. He first started writing about this in 1946, and in the 1950s worked with colleagues William Fowler and husband and wife team Margaret and Geoffrey Burbidge, to develop sophisticated theoretical models of how the elements were "cooked" in stars through the nuclear fusion process which was believed to power them (see e.g. [29]).

The big bang theory, as we saw in Sect. 3.4, was used by Alpher, Herman and Gamow to argue that the elements were created in the hot, dense early Universe. It turned out that neither theory was able to agree with the observations, which were that the Universe is 75 % hydrogen, 24 % helium and 1 % everything else. The big bang theory could successfully produce the amount of hydrogen and helium, but ran into problems with the heavier elements. The problem was that there were no known stable elements with atomic numbers 3, 4 and 5, so building carbon (atomic number 6) and heavier elements was impossible in the big bang theory, as the Universe would have cooled and expanded too much before Carbon could have been created from the proposed neutron-capture process.

On the other hand, the steady state theorists' argument that the elements were all created in stars also had a problem. Its problem was the reverse of the big bang theory's problem, it could successfully explain the 1 % of the heavier elements, but it was not able to explain the 24 % of helium, stars just did not produce that much helium. So, neither theory could completely explain the known abundances of the elements.

3.7.3 The Distribution of Matter in Space and Time

Ever since the days of Copernicus, it had been argued that the Earth was not in a preferred place. Initially it was not in a preferred place in the Solar System, then in our Galaxy, and when Hubble showed that the Universe was filled with galaxies all rushing away from ours, it was argued that they were, in fact, all rushing away from each other. This principle is called the *cosmological principle*, and it states that the Universe looks the same in all directions and that any part of it looks the same as any other part, or *homogeneous* and *isotropic* to use the scientific terms.

An additional homogeneity in the steady state theory would be the uniformity of the Universe with time. If one were to look back in time, the steady state theory predicted that the Universe would look the same as it does today. The big bang theory, on the other hand, predicted that a younger Universe would look different. Galaxies would be closer together, younger, and possibly different, and one would expect to find very few or no new galaxies in the Universe today. Because of this additional homogeneity of time in the steady state theory, it became known as the "*perfect cosmological principle*".

As things stood at the end of the 1940s and early 1950s, there were insufficient observations of the distribution of matter in space and time to argue one way or the other.

3.7.4 The Temperature of the Universe

As Alpher and Herman would show in their 1948 paper (see Sect. 3.8), the big bang theory predicted a hot beginning to the Universe, and a relic radiation from this hot beginning which should be at a temperature of about 5 K today. But, in his 1955 book *"Frontiers of Astronomy"* [34], Hoyle begins Chap. 19 with a section on observational tests of cosmology. He states

> Was there ever a superdense state?
> It is a suspicious feature of the explosion [big bang] theory that no obvious relics of a superdense state of the Universe can be found.

which he of course used as an argument against its veracity.

3.8 The First Prediction of the Cosmic Microwave Background

Building on the work Alpher had done in his Ph.D. thesis on the origin of the elements, he and Gamow and fellow APLJHU researcher Robert Herman worked on a number of the implications of an era of nucleosynthesis in a hot early Universe. Ever the prankster, Gamow tried to persuade Herman to change his name to "Delta", as he dreamed of publishing an "Alpher, Bethe, Gamow, Delta" paper! In October 1948 Gamow published a paper in *Nature* in which he summarised some of the findings of his group on the early Universe's nucleosynthesis [35]. However, Gamow was not good at detail, and made several calculation mistakes in the paper.

Alpher and Herman wrote a paper which appeared in *Nature* in December of the same year [36], correcting Gamow's mistakes. The corrections to Gamow's earlier letter is all that is mentioned in most of the short letter (see Fig. 3.5), but later in the two page letter Alpher and Herman make the first ever prediction of a remnant radiation from the hot early Universe, and predicted its present temperature to be *"about 5 K"* (highlighted in Fig. 3.5). They soon followed up this initial paper with another in *Physical Review* [37], making the same prediction of a remnant radiation from the early Universe which would have a temperature of *"about 5 K"*. In the period between 1948 and 1950 Alpher, Herman and Gamow published several dozen papers on nucleosynthesis in the early Universe, and the relic radiation left over from this period, refining their calculations as much as they could. The predicted temperature varied between versions of the calculations, but was always between about 5 and 50 K.

Fig. 3.5 A copy of Alpher and Herman's 1948 paper in *Nature* magazine which predicted the existence of a relic radiation left over from the hot early Universe, which I have *highlighted* in the *box*

Evolution of the Universe

In checking the results presented by Gamow in his recent article on "The Evolution of the Universe" [*Nature* of October 30, p. 680], we found that his expression for matter-density suffers from the following errors : (1) an error of not taking into account the magnetic moments in Eq. (7) for the capture cross-section, (2) an error in estimating the value of α by integrating the equations for deuteron formation (the use of an electronic analogue computer leads to $\alpha = 1$), and (3) an arithmetical error in evaluating ρ_0 from Eq. (9). In addition, the coefficient in Eq. (3) is 1·52 rather than 2·14. Correcting for these errors, we find

$$\rho_{mat.} = \frac{4 \cdot 83 \times 10^{-4}}{t^{3/2}}.$$

The condensation-mass obtained from this corrected density comes out not much different from Gamow's original estimate. However, the intersection point $\rho_{mat.} = \rho_{rad.}$ occurs at $t = 8\cdot6 \times 10^{17}$ sec. $\simeq 3 \times 10^{10}$ years (that is, about ten times the present age of the universe). This indicates that, in finding the intersection, one should not neglect the curvature term in the general equation of the expanding universe. In other words, the formation of condensations must have taken place when the expansion was becoming linear with time.

Accordingly, we have integrated analytically the exact expression[1] :

$$\frac{dl}{dt} = \left[\frac{8\pi G}{3} \left(\frac{aT^4}{c^2} + \rho_{mat.} \right) l^2 - \frac{c^2 l_0{}^2}{R_0{}^2} \right]^{1/2},$$

with $T \propto 1/l$ and $R_0 = 1\cdot9 \times 10^9 \sqrt{-1}$ light-years. The integrated values of $\rho_{mat.}$ and $\rho_{rad.}$ intersect at a reasonable time, namely, $3\cdot5 \times 10^{14}$ sec. $\simeq 10^7$ years, and the masses and radii of condensations at this time become, according to the Jeans' criterion, $M_c = 3\cdot8 \times 10^7$ sun masses, and $R_c = 1\cdot1 \times 10^3$ light-years. The temperature of the gas at the time of condensation was 600° K., and the temperature in the universe at the present time is found to be about 5° K.

We hope to publish the details of these calculations in the near future.

Our thanks are due to Dr. G. Gamow for the proposal of the topic and his constant encouragement during the process of error-hunting. We wish also to thank Dr. J. W. Follin, jun., for his kindness in performing the integrations required for the determination of α, on a Reeves Analogue Computer. The work described in this letter was supported by the United States Navy, Bureau of Ordnance, under Contract NOrd-7386.

Ralph A. Alpher
Robert Herman

Applied Physics Laboratory,
Johns Hopkins University,
Silver Spring, Maryland.
Oct. 25.

[1] Gamow, G., *Phys. Rev.*, **70**, 572 (1946).

3.9 From Where Does This Microwave Background Come?

In the early Universe, the temperatures are still too high for the electrons to combine with nuclei, so all of the matter is an ionised gas, what we call a *"plasma"*. The Sun is a good example of a plasma, it is made up of hydrogen and helium, but the temperatures are so hot that the electrons are too thermally excited to be associated with any of the nuclei. Electrons are very very good at scattering and absorbing photons, the particles of electromagnetic radiation. So good in fact that a photon which starts at the centre of the Sun will take as much as 170,000 years to reach the surface, as it goes on a random walk being constantly scattered and absorbed by the sea of free electrons. It is only when the plasma has become thin enough that the photons can escape, and this is at what we call the "surface" of the Sun. In reality, it is not a surface, it is just where the ball of gas which is the Sun has a low enough density for the photons to escape.

The photons in the early Universe had a similar problem, they were flying around in a sea of free protons, helium nuclei and electrons. The photons could not get anywhere without interacting with these electrons, and as a consequence the Universe was opaque, like a thick fog. But, as the Universe expanded and cooled, there came a time when the temperature had dropped enough for the protons and helium nuclei to be able to capture the free electrons. Theory shows that this should happen when the temperature had dropped to about 3,000 K, about the same temperature as the surface of a red giant star.

Astronomers are sometimes guilty of giving things misleading names, and this important moment in the evolution of the Universe is one of them—we sometimes call it "recombination", which implies that the protons and electrons are combing again. In fact, it was the first time that they combined, and the moment at which the Universe became neutral and the thick fog lifted. Another, better term given to this moment is "decoupling", as it is when the radiation and the matter decoupled from each other. From this moment, the radiation was finally free to escape, and it is the photons which decoupled from matter at this time which we now see in the microwave part of the spectrum. The photons have been travelling through space for billions of years, and have been redshifted into the infrared and microwave parts of the spectrum by the expansion of the Universe during the time between decoupling and now. In that time, the Universe has grown to be about one thousand times bigger, and this stretching of space has stretched the wavelength of the photons by the same factor, taking them from the visible part of the spectrum into a part which was invisible to us until well into the twentieth century.

3.10 A New Window on the Heavens

What we refer to as light is, in fact, just one part of so-called *"electromagnetic radiation"*, the part to which our eyes are sensitive. But there are many other types of electromagnetic (EM) radiation such as radio, microwave, infrared, X-rays and gamma rays. These names are familiar to most of us today, but none of them was known until 1800 when William Herschel accidentally discovered the infrared. The following year the ultraviolet was discovered, and by the late 1800s scientists were unravelling the mysteries of radio waves, and X-rays and gamma rays were discovered in the final few years of the 1890s. Each of these different parts of the EM spectrum has become useful in various ways. We use X-rays to look at our bones through our skin, we use microwaves to cook our food faster than conventional ovens, and of course we use radio waves to transmit television and radio programmes into our living rooms.

The visible part of the spectrum is actually a very tiny part of the whole EM spectrum. If the entire EM spectrum were represented by the keys of a piano, the bass keys (low frequency) would represent radio waves and the treble keys (high frequency) would represent the energetic X-rays and gamma rays. Representing the EM spectrum in this way, the visible part of the spectrum would correspond to less than one piano key! There are vast parts of the spectrum that we just did not know about nor know how to measure until early in the twentieth century.

We are all familiar with this use of radio waves as a way of listening to sound broadcasts. But, natural phenomena in space give rise to radio signals too. The first person to detect radio waves coming from space was Karl Jansky, who worked for the Bell Telephone Laboratories. In the early 1930s, Jansky was investigated the source of static that was interfering with short wave transatlantic voice communications. Using a large directional antenna, Jansky noticed that there was a signal which repeated every 24 h. Initially Jansky thought that the Sun was the source of this signal, but after more careful analysis he realised that the signal was not repeating every 24 h, but rather every 23 h and 56 min, the so-called *"sidereal day"*. This is the time a given star takes to cross a local meridian on two successive occasions, or to put it another way it is the time the Earth actually takes to spin 360° on its axis. The reason a "solar day" is longer is that the Earth has to turn an extra little bit for the Sun to cross the meridian, as we have moved about it in our orbit in the intervening time.

Finding that the celestial radio signal was crossing the meridian in sidereal time meant only one thing—the signal had to be coming from an object beyond our Solar System. After further analysis and comparison with star charts (Jansky was not an astronomer), he realised that the source of the signal was in the direction of Sagittarius, and therefore he realised that it was coming from the centre of the Milky Way galaxy. He announced his discovery in 1933 [38], and it even made the May the 5th 1933 edition of *The New York Times* [39].

Jansky wanted to follow up this discovery, and submitted a proposal to his bosses at Bell Labs to build a 30 m diameter radio dish. His proposal was, however, rejected

3.10 A New Window on the Heavens

as it was felt that the signal he had discovered would not impact on their planned transatlantic communication system, so Jansky was assigned onto a different project and never did any more radio astronomy. A few years later, radio engineer and keen amateur astronomer Grote Reber built his own radio telescope in his back garden in Illinois and conducted the first systematic survey of astronomical radio waves [40]. In honour of Jansky's pioneering work in radio astronomy, the unit of radio flux (energy per unit time per unit area per unit frequency interval) astronomers use is called the *Jansky*.

Although Jansky's pioneering work was important in showing that celestial objects could emit radio waves, it was two particular developments which lead to the explosion of radio astronomy in the late 1940s and early 1950s. The first was the Second World war, and the development of RADAR in Britain and the USA. The research into RADAR, which proved so vital in tracking the movement of enemy aeroplanes, led to the development of much higher sensitivity and much more compact radio receivers. After the war this new technology was used by scientists who were interested in astronomy to carry on the radio astronomy work that Jansky and Reber had started in the 1930s.

The second development was a theoretical one. As I've already stated, astronomers knew that some 75 % of the Universe was hydrogen. Some of this hydrogen could be detected in visible light because it gets excited by hot young stars and fluoresces. The Orion nebula is an example of this, it's an example of an "*emission nebula*". But, most of the hydrogen in the Universe was expected to be neutral not ionised, and astronomers had no way to detect it. No way, that is, until Dutch astronomer Hendrik van de Hulst made the prediction as part of his Ph.D. thesis in 1944 that neutral hydrogen would emit a very low energy photon at a wavelength of 21 cm [41], which is in the radio part of the spectrum. This comes about because the electron in its ground state can actually be in two different configurations. It can either be configured with its spin in the same direction as the spin of the proton in the nucleus, or with its spin in the opposite direction. There is a tiny energy different between the two states, and left on their own a hydrogen atom will flip from the higher to the lower energy state, and in doing so it emits a very low energy and hence very long wavelength photon.

After the second World war, a group of people who had worked on radar during the war and led by John Ratcliffe established a radiophysics group at Cambridge University. They started off by observing and studying the radio emission from the Sun. The group soon branched out into observing other sources, and Martin Ryle and Antony Hewish of the Cavendish Laboratory established the Mullard Radio Astronomy Group in the early 1950s. Meanwhile, Manchester University established Jodrell Bank Observatory, and in 1950 their 218-foot Transit Telescope made the first detection of extra-galactic radio waves when radio emission from the Andromeda galaxy was detected.

In the same year, 1950, Ryle and his colleagues published their first catalogue of radio sources, a catalogue which has come to be known as 1C (one-Cambridge) [42]. The observations were made at 3.7 cm using an antenna array (basically a field of antennae), and in this first survey the team discovered some fifty discrete radio sources. By 1955 the team had conducted a second survey (imaginatively known as the 2C survey!) [43], and in this survey they detected 1936 sources between a declination of $-38°$ and $+83°$. Of these, 500 of the most intense could have their positions determined to an accuracy of about $\pm 2'$ in Right Ascension, and about $\pm 12'$ in declination. The team found most of the sources were of small angular diameter, and were distributed isotropically over the sky (that is to say in all directions).

About 30 of the sources were of larger angular diameter, between 20 and 180', but the majority of these larger sources were close to the plane of the Milky Way galaxy and so the authors suggested that they represented a "rare class of galactic object". They then went on to say that about 100 of the sources appeared to be related to objects which were in the *New General Catalogue* [44] or the *Index Catalogue* [45, 46]; both optical catalogues of nebular objects which had been put together in the 1800s and the first decade of the 1900s. Importantly, they also found a distribution of signal strengths from these sources which suggested that there was a higher density of more distant radio sources than of nearby ones. This was the first piece of evidence that the properties of galaxies were different in the past (more distant galaxies) than in the present, which of course flies in the face of the steady state theory but is what one would expect in the big bang theory.

An even stronger piece of evidence of the evolution of the Universe came with analysis of the group's third survey, 3C. This survey was completed in 1959 [47], with a revised version produced in 1962 [48]. There were only 471 sources in the 3C catalogue, and 470 in the revised version, but importantly the positions and fluxes of the sources had been determined to a much better accuracy than in the 2C survey.

The first object in the 3C catalogue to which an optical counterpart was found was the object 3C 48, in 1960 by Thomas Matthews and Allan Sandage [49] (both of Caltech). Using radio interferometry to narrow down its position, and then subsequent direct optical photographs, they found that 3C 48 corresponded to a faint blue star-like object. However, when its spectrum was taken, it looked unlike the spectrum of any known star. First of all it contained emission lines (the spectra of stars usually show absorption lines), and in addition the emission lines were broad not narrow as is usually the case with blue stars. But, most puzzlingly, the pattern of lines did not seem to fit any pattern that astronomers had seen before.

By 1963 Matthews and Sandage had found three starlike counterparts to three sources in the 3C catalogue, and published their work a paper entitled "*Optical Identification of 3c 48, 3c 196, and 3c 286 with Stellar Objects*" [49]. The nature of the three sources was not known, but at least it seemed that they had been identified.

A breakthrough happened in 1962. One of the other 3C sources, 3C 273 (shown in Fig. 3.6), was predicted to pass behind the Moon on several occasions. Using the Parkes Radio Telescope in Australia, Cyril Hazard and John Bolton were able to make measurements [50] which allowed Caltech astronomer Maarten Schmidt to find its optical counterpart. Using the Mount Palomar 200-in. telescope, Schmidt obtained a spectrum of the star-like object. The spectrum was as confusing as that of 3C 48, he could see broad emission lines but was not able to identify them, the pattern just didn't seem to make any sense at all.

3.10 A New Window on the Heavens

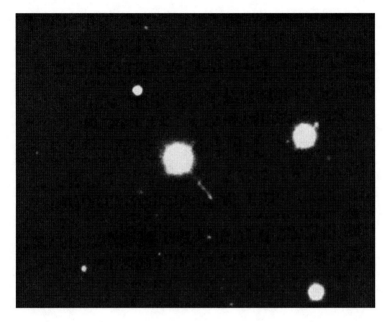

Fig. 3.6 An optical image of 3C 273 taken at Kitt Peak National Observatory. Image credit: NSF/Kitt Peak

After much head scratching and trying various things, Schmidt realised that the lines corresponded to hydrogen emission lines, but they were redshifted to such an extent that he had failed to recognise them. The redshift he measured for 3C 273 was nearly 16 % of the speed of light, an unheard of redshift at that time. Assuming Hubble's law which relates redshift to distance, this put 3C 273 at a huge distance from Earth, much further away than any galaxy ever seen. This work was published in a one page letter in *Nature* in 1963—"*3C 273: a star-like object with large redshift*" [51].

On the same day as his realisation that 3C 273's spectrum was highly redshifted, Schmidt bumped into his colleague Jesse Greenstein in the corridor of the Physics department at Caltech, and told him what he had found. Greenstein ran off to his office to start work on trying to identify the spectrum of 3C 48. By the end of the afternoon, Greenstein had identified the redshift of 3C 48, and found it to be 37 % of the speed of light, over twice that of 3C 273! His paper, also published in *Nature* in 1963, was entitled "*Red-Shift of the Unusual Radio Source 3C48*" [52]. The acronym "*quasar*" was coined by astrophysicist Hong-Yee Chiu. In 1964, he said in an article in *Physics Today* magazine [53]

> So far, the clumsily long name 'quasi-stellar radio sources' is used to describe these objects Because the nature of these objects is entirely unknown, it is hard to prepare a short, appropriate nomenclature for them so that their essential properties are obvious from their name. For convenience, the abbreviated form 'quasar' will be used throughout this paper.

As more research was conducted it was found that not all these quasi-stellar objects with extremely high redshifts had strong radio emission. Some were "radio quiet", so the acronym "*QSO*" (Quasi-Stellar Object) was coined, but because the term "quasar" had been in use for quite a while by this time, many or most astronomers refer to these objects as "quasars" whether they are radio-loud or radio-quiet. Throughout the 1960s the argument raged as to whether the observed Doppler shift of quasars was due to their being at a great distance (due to Hubble's law), or whether some other process was causing it. The steady state supporters argued that the redshift was being caused by e.g. gravitational effects, that as the photons escaped from an intense gravitational field they became redshifted in the process.

One of the problems with this suggestion was that for a star-like object to have such an intense gravitational field it would need to have a mass far in excess of the Hyashi limit, which sets a theoretical limit on how massive stars can be. A second problem was that the emission lines seen in quasars included so called "forbidden-lines", which are lines which can only be produced in low density regions such as nebulae. A point-like object could not also appear nebulous if it were local. However, those who argued that their redshifts were truly cosmological had to explain the prodigious amount of energy implied by their brightness and huge distances. There wasn't any physics known in the 1960s to explain such large energies.

This energy problem would be solved in the 1980s, in the same decade as the host galaxies of quasars were first observed with the Hubble Space Telescope, settling the argument of their nature once and for all. But, even by the late 1960s it was clear that there were more quasars with high redshifts than with low redshifts, and this was used by Schmidt and others to argue that they were more common in the past. The lowest known redshift of a quasar is 0.056, which is three times higher than the redshift of the Norma Cluster. To put this another way, there are no quasars anywhere in the local Universe. This would be truly puzzling if they existed today, but would not be at all puzzling if they were evidence of a phenomenon which existed in the earlier Universe but which does not exist today.

Even by the mid to late 1960s, if one accepted that their redshifts implied huge distances, they were one of the major pieces of evidence for an evolving Universe, and a nail in the coffin of the steady state theory. But, as mentioned in Sect. 3.7.4, in 1955 Hoyle was able to deride the big bang theory because of the lack of one piece of crucial evidence, the relic of its hot beginning. That elusive piece of evidence would be found in 1964 and would provide the fatal blow to Hoyle's theory.

3.11 Missed Opportunities

In the 1950s Alpher and Herman tried to persuade the radio astronomy community to look for this relic radiation from the early Universe that they had predicted, but no one was willing to take up the challenge. They were repeatedly told that it could not be done, that the radiation was too cold and there would be too much error. However,

3.11 Missed Opportunities

it would seem that in the Soviet Union there were physicists willing to give such observations a try. In 1964 Russian theoretical physicists Igor Novikov and A.G. Doroshkevich wrote a paper suggesting that the Cosmic Microwave Background Radiation (CMBR) which Alpher and Herman had predicted in 1948 could and should be observed [54]. In a popular astronomy book written by Novikov much later, he recalled that it had been observed by Russian scientists in 1957! [55]

In this book, Novikov states that in 1983 he received a phone call from Tigran Shmaonov of Russia's Institute of General Physics. Shmaonov told Novikov that whilst he was working as a postgraduate student in the 1950s, together with radio astronomers Sëmen Khaikin and N.L. Kaidanovsky using an antenna similar to the one with which Arno Penzias and Robert Wilson would discover the CMBR in 1964, Shmaonov measured radio waves at 3.2 cm coming from space. These results were reported in his 1957 Ph.D. thesis, and published in a Soviet journal. In this work, Shmaonov concluded that

> The absolute effective temperature of radiation background appears to be 4 ± 3 K.

Even earlier, in 1954, French astronomer Émile Le Roux of the Nançay Radio Observatory, whilst doing a sky survey at a wavelength of 33 cm, reported a near-isotropic background radiation at 3 K, ± 2 K. But La Roux did not realise the significance of what he had found and did not investigate it any further.

There were other missed opportunities, even earlier than this. In 1941 Walter Adams, who succeeded Hale as Director of Mount Wilson Observatory, published a paper on interstellar absorption lines that he and Theodore Dunham had observed with a spectrograph mounted to the coudé focus of the 100-in. [56]. In a paper published the same year, Canadian astronomer Andrew McKellar, working at the Dominion Astrophysical Observatory in British Columbia, identified some of these lines, including one as being due to a rotational transition in the molecule CN (cyanogen, part of the cyanide family) [57], and stated in the paper

> Also from Adams' results on the interstellar CN lines, it can be calculated that the 'Rotational' [his quotes] temperature of interstellar space is about 2K.

At the time, astronomers assumed that the temperature of interstellar space should be absolute zero (0 K), but no-one picked up on this result as being strange. In fact, calculations by George Field suggested that the excitation of the CN molecule could be due to collisions between the molecules [58]. It was only after Penzias and Wilson's discovery that astronomers Neville Wolf and George Field made the connection between the excitation of the CN molecule and the bath of cosmic radiation which was causing it [59].

Even at Bell Labs itself there were missed opportunities. In 1961 Edward Ohm was doing work with the Holmberg antenna, looking at sources of microwave noise. AT&T were bouncing telephone signals off of the Echo I balloon, but the reflected signals were very weak. In order to improve their chances of detecting the weak reflected signals, Bell Labs engineer Ohm was given the task in 1961 of trying to find all sources of excess noise in the Holmberg antenna's system which might contaminate measuring the actual signal from the Echo balloon. In a very thorough and systematic analysis he accounted for all of it except for a signal of about three degrees. He wrote up his findings in the *Bell System Technical Journal* [60], but he did not worry too much about this three degree excess as he felt it was small enough that it was within the uncertainties of what he was accounting for.

Two years later, in 1963, another Bell Labs engineer W.C. Jakes re-analysed the Holmberg horn antenna for noise, essentially re-doing what Ohm had done, but doing it even more thoroughly. Jakes concluded there really was about 2.5 degrees of noise that just could not be explained. He published his findings in another engineering journal, not the sort of place astrophysicists would think of looking [61]. In 1964, two Russian physicists A.G. Doroschkevich and I.D. Novikov came across Ohm's paper as they had read Alpher and Herman's 1948 prediction of a relic radiation, and were looking to see whether anyone in the USA had actually found it [54]. They accepted Ohm's explanation that the three degree excess was probably not significant within the errors. Doroschkevich and Novikov never saw Jake's paper, and it is interesting to speculate that, had they done so, they probably would have recognised that the excess Jake found was indeed the radiation predicted by Alpher and Herman.

3.12 Clearing Out the Pigeon Droppings

Arno Penzias and Robert Wilson were both radio astronomers who found themselves working at the Bell Research Labs in Holmdel, New Jersey. Penzias had been born into a Jewish family in Munich in 1933, but due to the rise of Naziism his family escaped to the USA in 1939 and settled in the Bronx part of New York City. Penzias was interested in physics from an early age, and majored in the subject at the City College of New York. He then went to Columbia University to study for a Ph.D. in radio astronomy. His advisor at Columbia was Charles Townes, who would later win a Nobel Prize for his invention of the maser. Penzias' Ph.D. project was to build a very sensitive radio receiver which would use Townes' maser as a key component.

Penzias had hoped to use this ultra-sensitive radio receiver to detect radio waves being emitted by the hydrogen gas which was believed to lie between galaxies, but he was unable to do so and finished his Ph.D. in 1961 with a feeling of failure. He left Columbia and took up a position at Bell Labs. There, he was given the freedom to conduct pure research for part of his time, and the rest of his time he was expected to undertake commercial research projects which tied in with the various projects in which Bell Labs were engaged. One of these projects was the Telstar project, the earliest communication satellite. Bell Labs were having problems pointing their antennae at the satellite; basically they couldn't easily find it in the sky as they didn't know exactly where they were pointing their antenna.

Penzias suggested the elegant solution of using the known position of a radio galaxy to calibrate the antenna direction, and by doing this know where they were pointing, and then to use this information to find the Telstar satellite. This was a wonderful combination of combining a pure research background with a practical use, and was why Bell Labs were prepared to employ people like Penzias with a "pure science" background.

By 1963, Penzias had been joined at Bell Labs by another radio astronomer, Robert Wilson. Wilson had grown up in Texas, and majored in physics at Rice University in Houston, and had then gone on to Caltech in 1957 to study for a Ph.D. Whilst at Caltech he did a course in cosmology which was taught by Cambridge University's Sir Fred Hoyle, who in the 1950s was a regular visitor to Caltech after his pioneering work on nucleosynthesis in stars that he'd done in the early 1950s with Caltech's Willy Fowler and Geoffrey and Margaret Burbidge. Wilson's thesis was also on radio astronomy, and after obtaining his Ph.D. he went to work at Bell Labs, partly attracted by its 6-m Holmdel "horn antenna" which was sited in Crawford Hill, near the Holmdel headquarters.

This antenna had originally been designed to detect signals from the Echo balloon satellite, which had been launched in 1960. This satellite was squeezed into a 66-cm diameter launch capsule, but once it was in space it inflated into a massive silver sphere, 30 m in diameter. This was large enough to bounce signals between an Earth-based transmitter and Earth-based receiver, but the receiver needed to be exceptionally sensitive and well-shielded from radiation from the ground to detect the faint signals bouncing off of Echo. The Holmdel antenna was constructed in 1959 to be the receiving detector for these bounced signals, and was 15 m long with a square aperture measuring 6 m on each side. Its unusual design had been optimised to pick up very faint signals from the high altitude Echo balloon, whilst minimising the extraneous signals from any terrestrial sources.

By 1963 Bell Labs were no longer involved in the Echo project, so the Holmberg antenna was lying idle. Because it was so sensitive and well shielded from the ground, the antenna was ideal for searching the sky for astronomical radio sources, a project Penzias and Wilson were keen to do in their "non-allocated" free research time. Before they could start their search, they needed to fully understand the behaviour of the antenna, or to "characterise it" as a scientist would call it. As they were planning to search for faint radio sources, one of the things they needed to fully understand was the "noise" the antenna had, that is the background signal when it was not pointing at a radio source. This background signal could be due to radio signals from the ground, local radio interference, or to electronic noise in the detection equipment.

In order to make sure they were detecting just noise, they pointed the antenna at a part of the sky where there were no known radio sources. When they did so, they detected a weak signal, much stronger than they expected. As they wanted to conduct the most sensitive survey yet of radio sources, they chose not to ignore this faint signal, but rather to find out its origin. The noise could have been due to interference from nearby, terrestrial radio sources, or even from nearby electrical equipment. Even a car's starter motor can cause radio interference. Penzias and Wilson wanted to check all of these possibilities. But, as they scanned the horizon in all directions they found no difference when their antenna was pointing towards New York City or towards the New Jersey countryside (Fig. 3.7).

Fig. 3.7 Arno Penzias and Robert Wilson in front of the Holmdel horn with which they accidentally discovered the cosmic microwave background radiation in 1964 (image credit—NASA)

They next investigated whether the noise was coming from the antenna itself. They checked every electronic component in their detection system, making sure all the wires were well shielded and all electrical contacts were good. Again, the noise did not go away. They even noticed a pair of pigeons were nesting in the horn of the antenna, and leaving a "white deposit" on the walls of the horn (if you've ever seen a pigeon fly around an office as I did once you know how much pigeons can defecate in even 10 of 15 min!). They removed the pigeons, taking them to a place 50 km away and releasing them, and then scrubbed the horn clean. Unfortunately the pigeons found their way back, and started dirtying the horn all over again, so Penzias captured them a second time but this time killed them. The low level signal was still there, everything they tried failed to get rid of it.

They spent the best part of a year trying to reduce the level of the noise, and although it was reduced somewhat it did not entirely go away. They had to accept that there was some component of their signal which seemed to be coming from space. It did not change no matter where they pointed the antenna, nor did it change with the time of the day nor the season. Something, somewhere was emitting radio waves at all times, but most puzzlingly it was coming from all parts of the sky with a uniform strength. It was as if the whole sky was filled with a *very faint* radio signal.

Towards the end of 1964 Penzias attended an astronomy conference in Montreal. During this conference he mentioned this "noise" problem with their horn antenna to cosmologist Bernard Burke, who was based at the time at the Department of Terrestrial Magnetism of the Carnegie Institute in Washington, DC. It was a fortuitous meeting, because a few months later Burke received a draft copy of a paper written by Robert Dicke and James Peebles of Princeton University in which they predicted that the Big Bang should have left an all-pervasive background radiation which today would be in the radio part of the spectrum.

3.12 Clearing Out the Pigeon Droppings

Robert Dicke had done his bachelors degree at Princeton, and then gone on to do a Ph.D. in nuclear physics at the University of Rochester, finishing his doctorate in 1939. During the second World war he worked at the Radiation Laboratory at MIT on the development of radar. During this time he designed an instrument which measured the power of microwave signals, which became known as the "*Dicke radiometer*". From observations Dicke made with his radiometer from the roof of the Radiation Laboratory in 1946, he argued that the blackbody temperature of the radiation from space was less than 20 K, but he did not specifically refer to a "background" radiation.

Soon after conducting this experiment, Dicke returned to Princeton where he was to work for the rest of his career. Dicke was the rare combination of a brilliant theoretician *and* experimentalist. In 1957 he proposed an alternative to Einstein's general theory of relativity, which led in 1961 to the Brans-Dicke theory of gravity. It was this work on an alternative theory of gravity which got Dicke thinking about cosmology and the early Universe. He became very interested in the idea of an oscillating Universe, one which goes through a never ending cycle of expansions and contractions. Dicke felt that, if the Universe did indeed "bounce", that it must get extremely hot when it was small so as to get rid of all the heavy elements that had been created in the stars in the previous expansion phase. In this scenario that Dicke was imagining, the heavy elements would all get broken down back to hydrogen in the intense heat as the Universe got smaller and smaller before bouncing back into a new expansion phase.

A group of Ph.D. students shared a room at Princeton, which was and still is common at many universities. Amongst these students were Peter Roll, James Peebles, Bruce Partridge, Mark Goldenberg and David Wilkinson. One day, Dicke walked into their office and later Wilkinson would recall that Dicke was thinking aloud, throwing ideas which had been percolating in his head out into the room. He said to the students who were there (some were his students, others were not) that he had been thinking about the idea of an oscillating Universe and that he felt that the Universe would have become extremely hot if it had bounced. According to Wilkinson,

> that is what Dicke talked about most of the time, he really liked the idea of a bouncing Universe!

Dicke went on to say in his thinking aloud that if the temperature had become so high as to get rid of all the heavy elements in the bounce, there should be some left over radiation. He continued to think aloud and said that, since the Universe would have expanded so much since then, that the radiation should probably now be in the microwave part of the spectrum. He then drew a blackbody curve on the blackboard in the room, and talked a little more about what might have gone on. Wilkinson recounted that this whole incident was very informal, just Dicke thinking aloud. This was apparently quite common, he was often wandering into the students' shared office thinking aloud. The students would sometimes think about Dicke's ideas if they caught their interest, but most of the ideas he came in with would just disappear by the next day. With this particular idea of a left over radiation from a hot, dense phase of the Universe, Dicke pointed out that the Dicke radiometer that he had developed in the war would be an ideal detector to look for such microwave radiation. But, Dicke also realised that the radiation would be coming from every part of the sky, which would present a problem.

Normally radio astronomers were able to measure the signal strength (radio astronomers call it a "temperature") of sources by looking at the difference between looking on-source and off-source. But, with radiation coming from everywhere, as this relic radiation would, there was no "off-source". The solution was to use what was known as an "absolute load", that is a source with a known flux, and then swap between observing this known artificial source and the all-pervasive relic radiation. This wasn't easy, but it could be done. As Dicke saw it, the need for an absolute load was the main obstacle to a successful detection of his proposed background radiation. He had every faith in his Dicke radiometer and felt it had sufficient sensitivity, but the difficulty in setting up an absolute load correctly was the main thing worrying Dicke during this first conversation. Wilkinson himself knew next to nothing about microwave instrumentation at this time, so any of Dicke's concerns passed over his head!

Dicke looked for a gifted theoretician to help him with the calculations, and Peebles was brought to his attention. Peebles was a Canadian who had studied initially at the University of Manitoba, and then had moved to the US to do his Ph.D. at Princeton. Peebles was working in the particle physics group at Princeton, but 1 day he had a chance conversation with fellow Manitoban graduate student Bob Moore, who was 1 year ahead of Peebles at Princeton. Moore told Peebles that he had been attending research talks being given by a member of the faculty with whom he was working, Bob Dicke, and that these talks were much more interesting that the work Peebles was doing. Peebles decided to see if Moore was correct, so he went along to hear one of Dicke's research talks, and was hooked. He decided to switch fields, and Dicke set him he task of calculating the conditions in the early Universe. To quote from an interview with Peebles [62]

> Dicke had the idea that not only was the Universe denser and hotter in its earliest moments, but that there should be a leftover radiation from this initial hot period.

Peebles set to work on the detailed calculations. He realised quite quickly that helium and a few other elements would be produced in great quantities in the very early Universe. Next he worked out how much helium would be produced, and his calculations showed it would be about 25 %. Peebles was not a cosmologist or an astrophysicist, so he did not know what the observed figure was, so was delighted to discover that his calculations were in perfect agreement with the observed abundance. Peebles and Dicke had solved one of the thorniest issues in cosmology, the question Alpher and Gamow had tried to answer in their "Alpher, Bethe, Gamow" paper [28]—from where had the elements come? Dicke and Peebles had shown that, whereas the work of Hoyle and colleagues had correctly shown that the heavier elements were forged in stellar interiors, the hydrogen and helium were created in the early Universe, when the temperatures were high enough for the nucleosynthesis to occur. They had, ironically, shown that Gamow's "big bang" theory was partly correct, and Hoyle's ideas on stellar cooking were also partly correct.

When Burke got the preprint from Dicke and Peebles he telephoned Penzias, wondering whether the "noise" he and Wilson had been unable to get rid of was, in fact, the very same radiation Dicke and Peebles were predicting in their paper. Penzias was intrigued, and after some thought realised that he and Wilson almost certainly had detected this primordial background radiation. He called Dicke, and told him of the signal he and Wilson had detected. Dicke was stunned, and a little peeved. He and his experimental colleagues David Wilkinson and Peter Roll had been designing an instrument to detect the radiation that he and Peebles had predicted, and were planning to put it on top of the Princeton physics departmental building. When Dicke answered the telephone he had Wilkinson and Roll in his office discussing some of the details of their planned instrument. He came off the telephone after his brief conversation with Penzias and said [63]

> Boys, we've been scooped!

He and his team visited Penzias and Wilson the following day, as Bell Labs is less than an hour's drive from Princeton. Once they had examined the pair's data Dicke knew indeed that Penzias and Wilson had discovered what we now call the "*Cosmic Microwave Background Radiation*" (CMBR).

The two teams published back to back articles in the *Astrophysical Journal* in the summer of 1965. In the first paper, entitled "*Cosmic Black-Body Radiation*" [64], Dicke, Peebles, Roll and Wilkinson outlined the theoretical arguments behind the prediction of a background radiation, and mentioned that they had constructed a radiometer to detect just such a signal, but had not yet obtained any results. They then go on to say [64]

> we recently learned that Penzias and Wilson (1965)[65] of the Bell Telephone Laboratories have observed background radiation at 7.3-cm wavelength. In attempting to eliminate (or account for) every contribution to the noise seen at the output of their receiver, they ended with a residual of 3.5 ± 1 K. Apparently this could only be due to radiation of unknown origin entering the antenna.

Dicke et al.'s paper (Fig. 3.8) is immediately followed by the experimental paper by Penzias and Wilson [65]. Considering that it announces one of the most important cosmological results in history, it probably has the most understated title of any scientific paper—"*A Measurement of Excess Antenna Temperature at 4080 Mc/s*" (Fig. 3.9). It doesn't exactly grab your attention does it? In the paper, Penzias and Wilson gave details of the signal they had detected, and state at the end of the abstract

> A possible explanation for the observed excess noise temperature is the one given by Dicke, Peebles, Roll and Wilkinson (1965)[64] in a companion letter in this issue.

COSMIC BLACK-BODY RADIATION*

One of the basic problems of cosmology is the singularity characteristic of the familiar cosmological solutions of Einstein's field equations. Also puzzling is the presence of matter in excess over antimatter in the universe, for baryons and leptons are thought to be conserved. Thus, in the framework of conventional theory we cannot understand the origin of matter or of the universe. We can distinguish three main attempts to deal with these problems.

1. The assumption of continuous creation (Bondi and Gold 1948; Hoyle 1948), which avoids the singularity by postulating a universe expanding for all time and a continuous but slow creation of new matter in the universe.

2. The assumption (Wheeler 1964) that the creation of new matter is intimately related to the existence of the singularity, and that the resolution of both paradoxes may be found in a proper quantum mechanical treatment of Einstein's field equations.

3. The assumption that the singularity results from a mathematical over-idealization,

* This research was supported in part by the National Science Foundation and the Office of Naval Research of the U.S. Navy.

Fig. 3.8 A copy of the beginning of the Dicke et al. paper in Astrophysical Journal [64] (image credit—American Astronomical Society/NASA ADS)

A MEASUREMENT OF EXCESS ANTENNA TEMPERATURE AT 4080 Mc/s

Measurements of the effective zenith noise temperature of the 20-foot horn-reflector antenna (Crawford, Hogg, and Hunt 1961) at the Crawford Hill Laboratory, Holmdel, New Jersey, at 4080 Mc/s have yielded a value about 3.5° K higher than expected. This excess temperature is, within the limits of our observations, isotropic, unpolarized, and

Fig. 3.9 A copy of the beginning of the Penzias and Wilson paper in Astrophysical Journal [65] (image credit—American Astronomical Society/NASA ADS)

3.13 Was It Really a Blackbody?

Exciting though Penzias and Wilson's discovery was, one has to remember that this microwave radiation had only been measured at one wavelength, 7.35 cm. To argue that it was due to radiation from a blackbody at a few Kelvin above absolute zero with only one data point was something most astronomers were not willing to accept. As even a school student knows, you can fit anything to one data point.

A second data point at a different wavelength was desperately needed, and in December 1965 Wilkinson and Roll obtained a measurement at 3.2 cm using the apparatus the Princeton team had been building on the roof of Guyot Hall at Princeton. This second data point was published in 1966 in *Physical Review Letters* [66]. By 1967 a few more data points had been determined by various groups, and were shown in a plot in an article in *Nature* in 1967 by T.F. Howell and J.R. Shakeshaft of the Cambridge radio astronomy group [67]. Measurements at eight different wavelengths had now been made, and the plot shown by Howell and Shakeshaft shows an excellent agreement to a 3 K blackbody curve.

Over the next decade more and more data points were added, so that in a review paper given to the International Astronomical Union (IAU) Symposium in 1976

3.13 Was It Really a Blackbody?

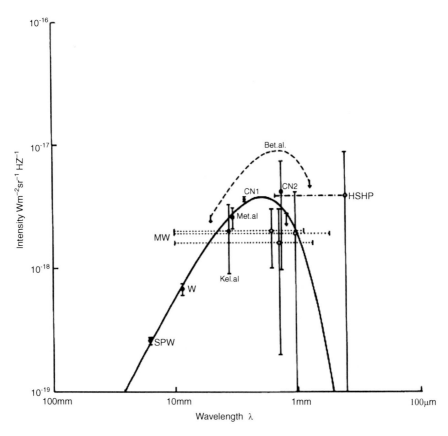

Fig. 3.10 From a review paper by E.I. Robson and P.E. Clegg and given by Robson at the IAU symposium in 1977 [68] (image credit—NASA)

held in Cambridge, Ian Robson and Peter Clegg of Queen Mary College, London were able to show the best fit to ten data points which had been gathered by various experiments since 1964 [68] (see Fig. 3.10).

The x-axis in Fig. 3.10 is the wavelength measured in millimetres, but note that (a) the scale is logarithmic and (b) it is *reversed* with longer wavelengths to the left. The data are well fitted by a blackbody curve with a temperature of 2.73 K, except for the errant data point denoted "HSHP" on the plot, which is from rocket-borne observations made between 5 μm (5 microns) and 1.3 mm by Houck, Soifer, Harwit and Pipher in 1972 from an altitude of 144 km above White Sands in New Mexico [69]. Using Wien's law, a blackbody with a temperature of 2.73 K peaks at a wavelength of 1.06 mm. On the Rayleigh-Jeans side of the curve (to the left of the peak in Fig. 3.10 as the wavelength increases to the left), one can see that the data points have small errors. But, on the Planck side of the curve (to the right of the peak) the error bars are much bigger, and there are very few of them. Although the data are well fitted by this blackbody curve, the data were not good enough to rule out other possibilities. Unfortunately for astronomers trying to measure the CMBR, the atmosphere glows like a furnace at wavelengths in the sub-millimetre and millimetre parts of the spectrum, due to the water vapour it contains. Astronomers realised they needed to get up above the atmosphere to make better measurements, particularly on the short wavelength side of the peak.

Although the arguments between the Steady State theory and the Big Bang theory would go on beyond Penzias and Wilson's 1965 discovery, the two were awarded the Nobel Prize in 1978. The citation from the Nobel committee was [70]

for their discovery of the cosmic microwave background radiation

Some people felt it was unfair that none of the theorists who had predicted its existence, Gamow or Dicke or even Peebles, was included (the Nobel Prize can be given to a maximum of three people). Gamow had died in 1968, and the Nobel Prize cannot be awarded posthumously, but why was Dicke not included with Penzias and Wilson?

The reason is probably that theory generally takes much more time to be accepted than experimental results. 1978 was only 13 years after Penzias and Wilson's discovery. A long enough time for the Nobel committee to be sure that the experiment was an important one and that it was telling us something very important about the nature of the Universe, but maybe not quite long enough for them to feel comfortable that the theoretical explanation of this radiation was completely correct. Overwhelming evidence that it was indeed due to a hot early Universe would come along by the early 1990s, with the launch of the Cosmic Background Explorer Satellite (COBE).

References

1. Einstein, A.: Die Grundlage der allgemeinen Relativitätstheorie (The basis of the general theory of relativity). Annalen der Physik **354**, 769–822 (1916)
2. Einstein, A.: Kosmologische Betrachtungen zur allgemeinen Relativitätstheorie (Cosmological Considerations on the General Theory of Relativity), pp. 142–152. Sitzungsberichte der Königlich Preußischen Akademie der Wissenschaften (Berlin), Seite (1917)
3. de Sitter, W.: Einstein's theory of gravitation and its astronomical consequences. Third paper. Mon. Not. R. Astron. Soc. **78**, 3–28 (1917)
4. Friedmann, A.: Über die Krümmung des Raumes (On the curvature of space). Zeitschrift für Physik **10**, 377–386 (1922). http://www.minkowskiinstitute.org
5. Friedmann, A.: The World as Space and Time. Minkowski Institute Press, Montreal (2014)
6. Friedmann, A.: Über die Möglichkeit einer Welt mit konstanter negativer Krümmung des Raumes (About the possibility of a world with constant negative curvature of space). Zeitschrift für Physik **21**, 326–332 (1924)
7. Lemaître, G.: Un Univers homogéne de masse constante et de rayon croissant rendant compte de la vitesse radiale des nébuleuses extra-galactiques (A homogeneous universe of constant mass and increasing radius accounting for the radial velocity of extragalactic nebulae). Annales de la Societe Scientifique de Bruxelles **47**, 49–59 (1927)
8. Lemaître, G.: Expansion of the universe, a homogeneous universe of constant mass and increasing radius accounting for the radial velocity of extra-galactic nebulae. Mon. Not. R. Astron. Soc. **91**, 483–490 (1931)

9. Lemaître, G.: Expansion of the universe, the expanding universe. Mon. Not. R. Astron. Soc. **91**, 490–501 (1931)
10. Einstein, A., & de Sitter, W.: On the relation between the expansion and the mean density of the universe. Contrib. Mt. Wilson Observatory **3**, 51–52 (1932)
11. Einstein, A., & de Sitter, W.: On the relation between the expansion and the mean density of the universe. Proc. Natl. Acad. Sci. USA **18**, 213–214 (1932)
12. Eddington, A.S.: The end of the world: from the standpoint of mathematical physics. Nature **127**, 447–453 (1931)
13. Bondi, H., Gold, T.: The steady-state theory of the expanding universe. Mon. Not. R. Astron. Soci. **108**, 252–270 (1948)
14. Hoyle, F.: A new model for the expanding universe. Mon. Not. R. Astron. Soc. **108**, 372–382 (1948)
15. Hoyle, F.: http://en.wikipedia.org/wiki/Fred_Hoyle#Rejection_of_the_Big_Bang. Retrieved 15 March 2014
16. Bryson, B.: A Short History of Nearly Everything. Black Swan, London (2003)
17. Becquerel, H.: Sur les radiations émises par phosphorescence (On the radiation emitted by phosphorescence). Proc. French Acad. Sci. **122**, 420–421 (1896)
18. Rutherford, E.: Collision of ? particles with light atoms, IV: an anomalous effect in nitrogen. Philos. Mag. **37**, 581–587 (1919)
19. Moore, R.: Niels Bohr: The Man, His Science & The World They Changed. Knopf, New York (1966)
20. Harkins, W.D.: The evolution of the elements and the stability of complex atoms. J. Am. Chem. Soc. **39**, 856 (1917)
21. Harkins, W.D.: The structure of atoms, and the evolution of the elements as related to the composition of the nuclei of atoms. Science **46**, 419 (1917)
22. Payne, C.H.: Stellar atmospheres; a contribution to the observational study of high temperature in the reversing layers of stars. Ph.D. thesis, Ratcliffe College (1925)
23. Bethe, H., Bacher, R.: Nuclear physics A. Stationary states of nuclei. Rev. Mod. Phys. **8**, 82–229 (1936)
24. Bethe, H.: Nuclear physics B. Nuclear dynamics. Theor. Rev. Mod. Phys. **9**, 69–244 (1937)
25. Eddington, A.S.: Stars and Atoms. Clarendon Press, Oxford (1927)
26. von Weizsäcker, C.F.: Über Elementumwandlungen im Innern der Sterne. II (On transformations of elements in the interiors of stars. II). Physikalische Zeitschrift **39**, 633–646 (1938)
27. Gamow, G.: Possible relation between deoxyribonucleic acid and protein structures. Nature **173**, 318 (1954)
28. Alpher, R.A., Bethe, H., Gamow, G.: The origin of chemical elements. Phys. Rev. **73**, 803–804 (1948)
29. Burbidge, E.M., Burbidge, G.R., Fowler, W.A., Hoyle, F.: Synthesis of the elements in stars. Rev. Mod. Phys. **29**, 547–650 (1957)
30. Herschel, W.: Experiments on the refrangibility of the invisible rays of the sun. Philos. Trans. R. Soc. Lond. **90**, 284–292 (1800)
31. Wien, W.: Die obere Grenze der Wellenlängen, welche in der Wärmestrahlung fester Körper vorkommen können; Folgerungen aus dem zweiten Hauptsatz der Wärmetheorie (The upper limit of the wavelengths which can occur in a solid body of the heat radiation; Consequences of the second law of thermodynamics). Annalen der Physik **285**, 633–641 (1893)
32. Tolman, R.C.: Relativity, Thermodynamics, and Cosmology. Clarendon Press, Oxford (1934)
33. Alpher, R.A., Herman, R., Gamow, G.A.: Thermonuclear reactions in the expanding universe. Phys. Rev. **74**, 1198–1199 (1948)
34. Hoyle, F.: Frontiers of Astronomy. Heinmann Educational Books, London (1955)
35. Gamow, G.: The Evolution of the Universe. Nature **162**, 680–682 (1948)
36. Alpher, R.A., Herman, R.: Evolution of the Universe. Nature **162**, 774–775 (1948)
37. Alpher, R.A., Herman, R.: Remarks on the evolution of the expanding universe. Phys. Rev. **75**, 1089–1095 (1949)
38. Jansky, K.G.: Radio waves from outside the solar system. Nature **132**, 66 (1933)

39. New radio waves traced to centre of the Milky Way. New York Times, 5 May 1933. http://en.wikipedia.org/wiki/May_1933#May_5.2C_1933_.28Friday.29. Retrieved March 2014 and http://daggy.name/cop/effluvia/jansky.htm. Retrieved March 2014
40. Reber, G.: Notes: cosmic static. Astrophys. J. **91**, 621–624 (1940)
41. van de Hulst, H.C.: Radio waves from space: origin of radiowaves. Ned. tijd. natuurkunde **11**, 210–221 (1945)
42. Ryle, M., Smith, F.G., Elsmore, B.: A preliminary survey of the radio stars in the Northern Hemisphere. Mon. Not. R. Astron. Soc. **110**, 508–523 (1950)
43. Shakeshaft, J.R., Ryle, M., Baldwin, J.E., Elsmore, B., Thomson, J.H.: A survey of radio sources between declinations $-38°$ and $+83°$. Mem. R. Astron. Soc. **67**, 106–154 (1955)
44. Dreyer, J.L.E.: A new general catalogue of nebulae and clusters of stars, being the catalogue of the late Sir John F.W. Herschel, Bart., revised, corrected, and enlarged. Mem. R. Astron. Soc. **49**, 1–237 (1888)
45. Dreyer, J.L.E.: Index catalogue of nebulae found in the years 1888 to 1894, with notes and corrections to the new general catalogue. Mem. R. Astron. Soc. **51**, 185–228 (1895)
46. Dreyer, J.L.E.: Second index catalogue of nebulae and clusters of stars; containing objects found in the years 1895 to 1907, with notes and corrections to the new general catalogue and to the index catalogue for 1888–94. Mem. R. Astron. Soc. **59**, 105–198 (1908)
47. Edge, D.O., Shakeshaft, J.R., McAdam, W.B., Baldwin, J.E., Archer, S.: A survey of radio sources at a frequency of 159 Mc/s. Mem. R. Astron. Soc. **68**, 37–60 (1959)
48. Bennett, A.S.: The preparation of the revised 3C catalogue of radio sources. Mon. Not. R. Astron. Soc. **125**, 75–86 (1962)
49. Matthews, T.A., Sandage, A.R.: Optical Identification of 3c 48, 3c 196, and 3c 286 with stellar objects. Astrophys. J. **138**, 30–56 (1963)
50. Hazard, C., Mackey, M.B., Shimmins, A.J.: Investigation of the radio source 3C 273 by the method of lunar occultations. Nature **197**, 1037–1039 (1963)
51. Schmidt, M.: 3C 273 : A star-like object with large red-shift. Nature **197**, 1040 (1963)
52. Greenstein, J.L.: Red-shift of the unusual radio source: 3C 48. Nature **197**, 1041–1042 (1963)
53. Hong-Yui, C.: Gravitational collapse. Phys. Today **17**, 21 (1964)
54. Doroshkevich, A.G., Novikov, I.D.: Mean density of radiation in metagalaxy and certain problems in relativistic cosmology. Sov. Phys. Dokl. **9**, 111–113 (1964)
55. Novikov, I.D.: Black Holes and the Universe. Cambridge University Press, Cambridge, England (1984)
56. Adams, W.S.: Some results with the COUDÉ spectrograph of the Mount Wilson Observatory. Astrophys. J. **93**, 11–23 (1941)
57. McKellar, A., Kan-Mitchell, J., Conti, Peter S.: Molecular lines from the lowest states of diatomic molecules composed of atoms probably present in interstellar space. Publ. Dominion Astrophys. Observatory **7**, 251–272 (1941)
58. Peebles, J.E., Page, L.A., Partridge, B.R.: Finding the Big Bang, p. 74. Cambridge University Press, Cambridge, England (2009)
59. Peebles, J.E., Page, L.A., Partridge, B.R.: Finding the Big Bang, pp. 77–83. Cambridge University Press, Cambridge, England (2009)
60. Lemonick, M.D.: Echo of the Big Bang, p. 38. Princeton University Press, Princeton (2003)
61. Jakes, W.C.: Participation of the Holmdel Station in the Telstar Project, Telstar I. NASA SP-32, pp. 1421–1447. NASA, Washington
62. Harwitt, M.: Interview with Dr. James Edwin Peebles. (1984). http://www.aip.org/history/ohilist/4814.html
63. Harwitt, M.: Interview with Dr. David Wilkinson. (1984). http://www.aip.org/history/ohilist/4967.html
64. Dicke, R.H., Peebles, P.J.E., Roll, P.G., Wilkinson, D.T.: Cosmic black-body radiation. Astrophys. J. **142**, 414–419 (1965)
65. Penzias, A.A., Wilson, R.W.: A measurement of excess antenna temperature at 4080 Mc/s. Astrophys. J. **142**, 419–421 (1965)

References

66. Roll, P.G., Wilkinson, D.T.: Cosmic background radiation at 3.2 cm-support for cosmic blackbody radiation. Phys. Rev. Lett. **16**, 405–407 (1966)
67. Howell, T.F., Shakeshaft, J.R.: Spectrum of the 3° K cosmic microwave radiation. Nature **216**, 753–754 (1967)
68. Robson, E.I., Clegg, P.E.: Microwave background spectrum - survey of recent results. In: Jauncey, D.L. (ed.) Radio Astronomy and Cosmology; Proceedings of the Symposium, Cambridge University, Cambridge, 16–20 August 1976. IAU Symposium, No. 74, pp. 319–325 (1977)
69. Houck, J.R., Soifer, B.T., Harwit, M., Pipher, J.L.: The far-infrared and submillimeter background. Astrophys. J. **178**, L29–33 (1972)
70. The Nobel Prize in Physics. http://www.nobelprize.org/nobel_prizes/physics/laureates/1978 (1978)

Chapter 4
The Cosmic Background Explorer (COBE)

In 1977 Smoot and colleagues at Berkeley discovered that the motion of our Galaxy through space is 625 km/s, much larger than anyone had expected. Around the same time, Rubin and Bosma independently found evidence for "dark matter", rediscovering a finding that Zwicky had made in 1933. In the late 1980s Geller and Huchra find the distribution of clusters of galaxies to be highly non-uniform, with great filaments and large voids. The Cosmic Background Explorer (COBE) satellite was launched in 1989, and in 1990 it showed that the cosmic microwave background was a *perfect* blackbody. In 1992 COBE found the long-sought after fluctuations in the cosmic microwave background, the seeds of the structure that we see in the Universe today. These two discoveries by COBE led to John Mather and George Smoot being awarded the 2006 Nobel Prize in Physics.

4.1 The Earth, Sun and Milky Way's Motions in Space

In 1967 Dennis Sciama realised that, if the signal detected by Penzias and Wilson a few years before was indeed cosmic in origin, then the Earth's motion in space should lead to it looking slightly different in different parts of the sky [1]. Technically this is called a 'dipole moment', and is simply due to the same Doppler shift we have talked about already. The Earth orbits the Sun, and the Sun orbits the centre of the Milky Way, and when you add these motions together the Earth is moving at about 200 km/s in space. Most of this 200 km/s is due to the Sun's motion around the centre of the Milky Way. Although 200 km/s is small compared to the speed of microwaves (which travel at the speed of light), the dipole moment is an effect which should have been measurable. In the late 1960s, Wilkinson and graduate student Bruce Partridge adapted the antenna on the roof of Princeton's geology building and spent a year mapping the sky looking for it. Their paper "Isotropy and Homogeneity of the Universe from measurements of the Cosmic

Microwave Background" was published in 1967 [2], reporting that the cosmic microwave background radiation (CMBR) was smooth to better than 0.1 % of 3 K. Wilkinson recalled later that their measurements hinted that there was something there, but the Princeton team didn't feel the result was unambiguous enough to say they had found it.

Peebles repeated Sciama's suggestion of a dipole moment in his 1971 book *'Physical Cosmology'*, a book which became the blueprint for experimentalists wanting to see what tests could be done of the big bang theory [3]. Astronomers realised that the small change in the signal that they were looking for was being overwhelmed by the emission in the microwave part of the spectrum from the atmosphere and the only way to do the experiment properly was to get above the atmosphere. Unfortunately, space missions take a long time to plan and put into implementation, well over a decade. In 1974 plans started to be put together for a satellite which would investigate the cosmic microwave background to a level of detail which was just not possible from the ground—the mission would become known as the Cosmic Background Explorer (COBE), and would eventually be launched in November 1989.

In the meantime, astronomers did they best that they could. There was a race on to measure the dipole moment, with several teams vying to be the first. One such team was based at the Lawrence Berkley National Laboratories (LBNL) in California, a small team with George Smoot as one of its most enthusiastic members (Fig. 4.1). They knew that such an experiment could only work if one could get above as much of the water vapour in the atmosphere as possible and so they considered White Mountain in California, which was at an altitude of 14,000 ft and very dry at the summit. Smoot knew that putting the experiment in a high-altitude balloon would be even better, but after his experiences of working with such balloons in experiments looking for anti-matter, he was not keen to continue working with such a temperamental observing platform. But then, in a fortuitous twist of fate, Smoot happened to attend a talk in 1973 being given by Townes who, remember, had been Penzias' Ph.D. supervisor, in which Townes was describing the Kuiper Airborne Observatory (KAO), which was going to begin operation the following year. This is the same observatory that I flew in several times in 1995 when I began my career in airborne astronomy at the University of Chicago.

Smoot had started his research career as a particle physicist. After doing his bachelors degree in mathematics and physics at MIT, he had stayed there to do a Ph.D. in particle physics. He then went to work with Luis Alvarez at the LBNL on an experiment to look for anti-matter using a detector in a high-altitude balloon. After 4 years of experiments, they failed to find any anti-matter. Soon after Smoot had heard Townes' talk about the KAO, Alvarez advised his research team to consider switching to another research area rather than run the risk of staying working on this possibly fruitless search for anti-matter.

By now, Smoot was approaching 30 years of age, and decided to take Alvarez's wise advice. He started thinking about some other interesting cosmological experiments in which he could get involved, and started studying Peebles' 1971 book *Physical Cosmology* in far more detail than he had done before. He fell upon the part

4.1 The Earth, Sun and Milky Way's Motions in Space

Fig. 4.1 George Smoot (image credit: Michael Hoefner)

of the book where Peebles had talked about the '*aether drift*' experiment (looking for a dipole in the CMBR), and decided that it was something he wanted to pursue.

Smoot realised that doing the dipole experiment from an aeroplane had a lot of advantages over doing it using a balloon, the main one being that the experimenters could fly with the apparatus, fix any problems in flight if necessary, and the equipment had less chance of being destroyed upon landing. He started exploring this possibility, but the KAO that NASA was about to put into operation was a C-141 military transport plane and did not fly high enough to get above enough of the Earth's atmosphere. Then, Smoot and his boss Alvarez realised that a U2 spy plane could go much higher than the KAO. These were the same U2 aeroplanes that had been used to fly over Cuba in the early 1960s and had seen the installation of nuclear missiles in what became known as the Cuban missile crisis. The technical challenges of using a U2 plane were formidable; the space available for an experiment was tiny, requiring them to custom build their radiometer to fit in such a confined space. Also, they needed to build a radiometer with unprecedented accuracy, as Partridge and Wilkinson had already shown in 1967 that the effect they were looking for was less than 0.1 % of the CMBR.

After nearly 2 years of building equipment, testing, fixing problems and making necessary improvements, by late 1976 they were making U2 flights and the results started to come in. In December of that year Smoot and his colleague Jon Aymon were working night and day analysing the results from their latest flights. As they worked their way through the data, it became more and more evident that they had indeed found a dipole, the strength of the CMB was not the same in opposite parts of the sky. But, when they worked out the direction in which it was hotter (bluer) they were totally confused. The direction did not correspond to the direction in which the Milky Way is rotating! In fact, the sky was warmest in the direction of the constellation Leo, and coolest in the direction of Aquarius, meaning the Earth was moving towards the former and away from the latter. If the result was correct it could only mean one thing—that the Earth and the Sun and the whole Milky Way galaxy were moving in the direction of Leo. And, not just moving, but moving very fast, at about 600 km/s, which is more than a million miles per hour!

The result was sufficiently shocking that they needed to make sure they had not made a mistake. First they checked all their measurements, everything was fine. But, they were still worried that maybe it was a seasonal effect due to the Earth's motion about the Sun, so they scrambled to do a whole further series of U2 flights and analyse the data from them. They did so, and the result still stood, they were confident it was real and not a seasonal effect.

Smoot was able to announce this shocking result in the April 1977 meeting of the American Physical Society [4]. Because they had been working on analysing their data right up to the meeting, they did not actually have a formal slot in which to present their results. But, Peebles was scheduled to give an invited talk on cosmology, and he generously allowed Smoot to have a few minutes at the end of his own allotted time. Smoot presented the results, and when he handed back to Peebles the comment from Peebles was that the results from Smoot's team *"presented a real dilemma for theorists."* [4].

But, the announcement didn't really cause much of a stir, the audience were mainly physicists not astronomers or cosmologists. Some 6 months later Marc Gorenstein of the LBNL team presented the same results at the American Astronomical Society meeting in Atlanta [5], and this time it caught the attention of Walter Sullivan of the New York Times, who ran a front page headline

Galaxy's speed through Universe found to exceed a million m.p.h.

The only reasonable explanation for this motion was that something pretty massive in the direction of Leo is attracting our Galaxy, which would not happen if galaxies were distributed uniformly in space. This motion towards Leo therefore implied that matter in the Universe was not as uniformly distributed as the homogeneity in the CMBR implied. It implied that the Universe was clumpy, but if it was clumpy now then that clumpiness had to have been there in the early Universe, and should show up in the CMBR. The race was now on to do an experiment to try to observe this clumpiness.

However, it wasn't long before some astronomers suggested that Smoot and his colleagues had not really found a dipole, and that their results was a false one caused by only making their observations in the northern hemisphere. In order to counter this objection, they had to arrange to make a series of U2 flights in the southern hemisphere. Ideally they would have chosen Australia, but the U2 could not fly that far and would have to be transported bit by bit in a transport plane, which made the option too expensive. South America seemed like the best choice, but Chile and Argentina were threatening to go to war against each other at the time, so eventually the team chose Peru. After conducting a series of flights there and analysing the data they were relieved to find the 600 km/s measurement was still there, and so could convincingly argue that it was real and not due to a local effect. The strange mass which was causing our Galaxy to hurtle at 600 km/s in the direction of Leo was dubbed *'the great attractor'*, but what the heck was it?

4.2 Invisible Matter

One of the main objections to the 600 km/s motion found by Smoot and his colleagues was that there just didn't seem to be enough visible material in the direction of Leo to be able to pull our Galaxy towards it at such a high speed. However, evidence was building that astronomers were not seeing all the matter in the Universe no matter how sensitive their telescopes. There seemed to be a large amount of unseen matter which astronomers were arguing did not emit electromagnetic radiation at any wavelength, and they dubbed it 'dark matter'.

The idea of dark matter had actually been around since the 1930s, but like most ideas ahead of its time, it had been largely ignored. The discovery was made by Fritz Zwicky (Fig. 4.2), who at the time was working at Caltech and had access to the 100-in. at Mount Wilson. Zwicky had been born in Bulgaria, but when he was six he moved to Switzerland to live with his grandparents. In 1925, at the age of 27, he emigrated to the US to work at Caltech with Robert Millikan who had been awarded the Nobel prize for physics a few years earlier. Zwicky's office was just down the hallway from Robert Oppenheimer, who would go on to head up the Manhattan project to build the atomic bomb.

Zwicky quickly established himself amongst his colleagues as being not without talent but also in possession of an abrasive (some would say psychotic) personality. Most of his colleagues avoided him if they could. One of his closest collaborators, Walter Baade, an affable astronomer originally from Germany, was so intimidated by Zwicky's aggressive style that he refused to be left alone with him during their frequent nights together in the dome of the 100-in. The pair of them suggested the term "supernova" in the 1934 to classify a type of bright new star that was different from a normal "nova". They argued that a supernova was the transition of a normal star into a neutron star, a suggestion which is indeed correct.

Fig. 4.2 Swiss-American astronomer Fritz Zwicky who discovered dark matter and coined the term 'supernova'

Zwicky was often not very complimentary about his colleagues, describing them as "scatterbrains", "sycophants and plain thieves". He was quoted as saying that such people

> "have no love for any of the lone wolves who are not fawners and apple polishers ….. who doctor their observational data to hide their shortcomings and to make the majority of the astronomers accept and believe in some of their most prejudicial and erroneous presentations and interpretations of facts" who therefore publish "useless trash in the bulging astronomical journals."

In 1933, Zwicky turned his attention to studying the Coma cluster, and in particular the motions of the galaxies within the cluster. Although Slipher and Hubble had shown that most galaxies are moving away from us (see Chap. 2), galaxies within clusters also move about the centre of the cluster, like bees swarming about their central hive. By subtracting away the redshift of the cluster as a whole, Zwicky was able to study the speed of motion of the Coma cluster galaxies about the cluster's centre.

What he found was quite a shock, the galaxies in the Coma cluster were moving so fast that the cluster should fly apart, there just didn't seem to be enough material he could see in the galaxies to be able to hold them in their orbits. He postulated that there was a vast reservoir of unseen material, and in 1933 he coined the term *dunke Materie* ('dark matter') in a paper which appeared in *Helvetica Physica Acta*, a publication of the Swiss Physical Society [6]. In 1937 he published his studies of the Coma cluster in an English language paper entitled *"On the Masses of Nebulae and of Clusters of Nebulae"* [7], which appeared in the *Astrophysical Journal*, but despite the prominence of this journal, the work seems to have been largely ignored at the time. But, some 40 years later, other lines of evidence suggested that Zwicky had been right all along.

Firstly, on the theoretical side, in 1973 Peebles and fellow Princeton theoretician Jeremiah Ostriker wrote an influential paper entitled *"A Numerical Study of the Stability of Flattened Galaxies: or, can Cold Galaxies Survive?"* [8], suggesting that for typical galaxies there was insufficient visible matter in them to prevent their disks from flying apart. Their paper used computational models of simulated galaxies with 150–500 mass points, and followed their evolution. Their models started with disks that had characteristics similar to those observed for our Milky Way galaxy, but they found that the disks became unstable unless the galaxies were surrounded by a massive, unseen halo with as much or more mass than that of the disk. Only by adding these unseen halos could Ostriker and Peebles keep their model galaxy disks stable.

Later in the 1970s several astronomers, including Vera Rubin and her colleague Kent Ford in the US and Albert Bosma (working on his Ph.D. Thesis at the University of Groningen in The Netherlands), noticed something strange about spiral galaxies. Rubin had developed a habit of doing research that was both groundbreaking and years ahead of its time. After graduating from Vasser College, she had done her Masters degree at Cornell University, and in December 1950 presented the findings of her masters research at a meeting of the American Astronomical Society. In this work, she had found evidence for significant deviations from the uniform motion of galaxies due to the expansion of the Universe; we call this uniform motion the *'Hubble flow'*.

Rubin found that many galaxies showed motions in space very different from this Hubble flow, but the work was rejected by all the astronomical journals at the time so never appeared in the literature. After completing her masters degree at Cornell, Rubin went to GWU to do her Ph.D. under Gamow, and found in 1954 the first evidence for clustering of clusters of galaxies (what we now call *'superclusters'*) [9]. The only astronomer to really take Rubin's work seriously was French astronomer Gérard de Vaucouleurs. Working at the University of Texas, de Vaucoulerus gathered data to support the idea that clusters of galaxies were further grouped into superclusters, and de Vaucouleurs concluded that space was a lot less homogeneous than had been previously thought [10]. It turned out that this non-uniform distribution of galaxies was the reason for the 600 km/s motion of our Milky Way galaxy towards Leo that Smoot and colleagues were later to report from their U2 observations of the dipole in the CMB in 1977.

After completing her Ph.D., Rubin took a career break to bring up her family, and re-entered astronomy in the early 1960s. She became an assistant professor at

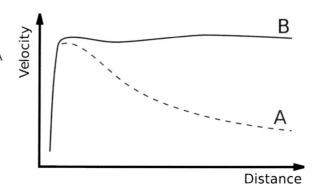

Fig. 4.3 The evidence for dark matter within galaxies. The rotation curves of spiral galaxies should follow line A if most of the matter is concentrated at a galaxy's centre. What we observe is line B, suggesting that most of a galaxy's mass is in a large halo surrounding the galaxy

GWU and then moved onto the staff at the Department of Terrestrial Magnetism at the Carnegie Institution in Washington. It was here that she started collaborating with Kent Ford and, using an image intensifier that Ford had developed, they started studying the motions of stars in the disks of galaxies by observing the Doppler shift in the stars' spectral lines.

As we have already mentioned, the stars in a spiral galaxy rotate about the centre, our Sun is moving about the centre of the Milky Way at about 200 km/s, and so will take about 250 million years to make one complete orbit. All spiral galaxies show this, but when Rubin and Ford [11, 12] and separately Bosma [13, 14], studied the motions of stars in different spiral galaxies they noticed something very remarkable. Most of the visible stars in a spiral galaxy lie towards its centre, and so as you look further and further away from the centres of galaxies the stars should move slower and slower in their orbits, exactly as happens in our Solar System with the speed of the planets. This is just Kepler's third law, which I mentioned in Chap. 1.

This is not what Rubin, Ford and Bosma found. Instead, they found that the speed of stars further and further from the centres of spiral galaxies was the same as for stars closer to the centre—see Fig. 4.3. The technical term astronomers use for this is that the rotation curves of spiral galaxies were *'flat'*. The explanation which is accepted by most astronomers today is that the flat rotation curves of spiral galaxies are due to their being surrounded by a massive halo. For a while some astronomers argued that the massive halo could be due to unseen, normal matter in the form of faint stars, neutron stars, white dwarfs and black holes. But extensive searches for such objects in the 1980s and early 1990s failed to find these in sufficient quantities, and by the mid 1990s most astronomers not only accepted the reality of massive halos, but conceded that the material which made up the halos was 'dark matter'. We still do not know what this dark matter is, but there are now several lines of evidence that it exists and is not made up of the normal protons and neutrons which make up normal matter, what we call *'baryonic matter'*.

4.3 Filaments and Voids

The 600 km/s motion of our Milky Way galaxy towards Leo that Smoot and his team announced in 1977 was one of the first hints that matter was far from uniformly distributed in space. Rubin's masters thesis work was another hint. Much more extensive evidence would emerge during the 1980s from work being done by a team at the Harvard Smithsonian Center for Astrophysics (CfA). In 1977, a team led by Margaret Geller and John Huchra of the CfA started a project which would become known as the 'CfA Redshift Survey'. This was a project to observe thousands of galaxies to record their position in the two dimensional sky, but crucially to also observe each galaxy's redshift so that its position in three dimensional space could be determined.

In 1982 the team published the results of their first survey, a map of a slice of the sky between a declination of $+26.5°$ and $+32.5°$ and between a right ascension of 8 and 17 h [15, 16]—see Fig. 4.4. In this slice, they had measured the positions and redshifts of all the galaxies brighter than an apparent magnitude of 15.5, about 1,100 galaxies in total. The redshifts ranged from the low redshift of nearby galaxies out to redshifts of 15,000 km/s for the most distant galaxies in their survey.

The results were quite a surprise to the astronomical community. As Fig. 4.4 shows, the distribution of galaxies is far from uniform. Instead, galaxies were found concentrated in what became known as *'filaments'*, with vast unoccupied *'voids'* lying between the filaments. A second larger survey was started in 1985 and continued until 1995 [17, 18]. This larger survey eventually measured the positions and redshifts of some 18,000 galaxies, ten times the number in the initial survey. Although the two dimensional positions of many galaxies could be obtained simultaneously, the redshift of each galaxy had to obtained individually using a spectrograph with a slit placed over each object, a painstaking and time consuming process. In Fig. 4.5 the result of this second survey is shown, confirming the filaments and voids that had been seen in the results from the first survey.

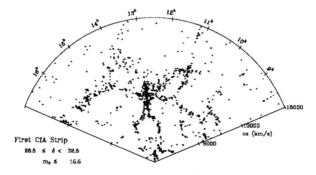

Fig. 4.4 The results of the first CfA redshift survey, published in 1982 (image credit: Smithsonian astrophysical observatory and Huchra et al. [16])

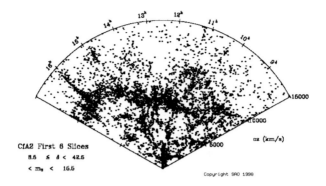

Fig. 4.5 The results of the second CfA redshift survey of about 18,000 galaxies (image credit: Smithsonian astrophysical observatory and Huchra and Geller 1998)

As part of the second survey, in 1989 Geller and Huchra announced the discovery of a vast concentration of galaxies stretching from a right ascension of 8–17 h and from a redshifts of 5,000–10,000 km/s, which translates to a structure about 600 by 250 by 30 million light years in size, by far the largest cosmic structure known at that time [19]. The object was given the name the *'Great Wall'*, yet more evidence that the distribution of clusters of galaxies was far from uniform, even on these large scales.

4.4 Cosmic Inflation

By the late 1960s theoreticians were becoming concerned about a number of aspects of the Universe. The three most troublesome problems were

1. the 'horizon' problem
2. the 'flatness' problem
3. the 'magnetic monopole' problem

It would be a young particle physicist Alan Guth who, in 1980, would come up with a very elegant solution to these three problems when he invented the idea of *'cosmic inflation'*.

4.4.1 The Horizon Problem

When Partridge and Wilkinson found in 1967 that the CMBR varied by less then 0.1 % of its ≈3 K temperature anywhere in the sky[2], this finding actually presented a problem. If one is looking in opposite directions in the sky one would not expect the two patches of the sky to have the same CMBR temperature. The CMBR was produced when the Universe was about 350,000 years old. Even though we are talking about photons, which obviously travel at the speed of light, 350,000 years was not enough time for photons in opposite parts of the sky to communicate their energy (and hence temperature) with each other.

This is a little like heating a cold room with a heater placed at its centre. We all know that it takes time for the whole room to warm up, because the heat will take time to travel around the room. Let us suppose it is quite a large room, 5 m by 5 m by 2 m in height, and yet we find that just 30 s after turning on the heater the whole room was at the same temperature, including the corners. This would be very strange, as the heat would not have had time to reach all parts of the room in this short time.

The same is true for opposite parts of the sky. Why should they be at the same temperature when the CMB was produced? Even when the Universe was 350,000 years old, it was too large for photons in opposite parts of the sky to have traversed it.

4.4.2 The Flatness Problem

As we have already discussed, Einstein's theory of General Relativity relates gravity to a distortion of space. Put simply, if the Universe has a lot of mass in it, it will have a positive curvature like a ball. If it has very little mass in it, it will have negative curvature like the surface of a saddle. And, if the amount of mass density is just right then it will have a flat geometry like a flat surface (see Fig. 4.6). The density required for a flat Universe is called the *'critical density'*. The ratio of the actual density to the critical density is called Ω. We often write this as Ω_0, the total Ω due to adding the contributions from different components such as normal matter, dark matter and anything else. The three possible scenarios are shown in Fig. 4.6.

The Universe could have any density, and yet we find that when we just include the normal baryonic matter, the density comes out to be about 4–5 % of the critical density. That is, Ω_b (the density of baryonic matter) is about 0.04–0.05. If we add in the dark matter, the density due to both normal and dark matter, Ω_M, comes out to be about 0.25. If it is within a factor of four to being unity now, then that means that it needed to be within a factor of about 0.0000000000000000000000000000000000000000000000000000000000001 (10^{-60} or one divided by ten followed by 60 zeros) in the very early Universe. To be this close to unity has led many theorists to argue that it *has* to be unity. But why?

4.4.3 The Magnetic Monopole Problem

In electricity we find both negative and positive charges. James Clerk Maxwell showed in the mid 1800s that electricity and magnetism are both part of the same force—electromagnetism. Therefore, in the same way that we find isolated positive and negative electric charges, we should find magnetic monopoles, naked 'north' or 'south' magnetic poles. But we don't, all magnets have both poles, we have never seen one without the other.

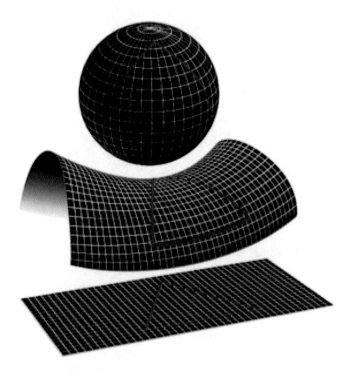

Fig. 4.6 The three possible geometries of the Universe, which depend in the ratio of its density to the so-called *'critical density'* Ω_0. If $\Omega_0 = 1$ (*bottom diagram*) the Universe is flat, if $\Omega_0 < 1$ (*middle diagram*) it is open, and if $\Omega_0 > 1$ (*top diagram*) it is closed

4.4.4 Cosmic Inflation Provides an Elegant Solution

Alan Guth (Fig. 4.7) started thinking about cosmology after hearing a talk by Dicke in 1979. As a particle physicist, he was already aware of the magnetic monopole problem, but Dicke's talk made him aware of the horizon problem and the flatness problem too. Guth was born in New Brunswick in New Jersey State, and graduated from MIT in 1968 with a bachelors degree in physics. He stayed there to do his masters and Ph.D., which he completed in 1971. From MIT he went to Princeton for 3 years, then to Columbia for 3 years, then to Cornell for 2 years before spending a year at the Stanford Linear Accelerator (SLAC), part of Stanford University. Being part of the 'baby boom' generation, he was finding it difficult to find a permanent academic job as there were too many well qualified young physicists around and too few older physicists retiring. He gave his first seminar on his ideas of cosmic inflation in January 1980, and this helped him obtain a tenure-track faculty position.

As Guth started thinking about a solution to the magnetic monopole problem, he realised that it could solve the flatness and horizon problems too. Like a lot of

4.4 Cosmic Inflation

Fig. 4.7 Alan Guth, who proposed the idea of cosmic inflation in 1980 (image credit: Betsy Divine)

very good ideas, Guth's idea of cosmic inflation is very elegant in its simplicity. The basic idea Guth put forward is that the Universe went through a brief period of *extremely rapid* expansion very early in its history, when it was about 10^{-36} s old (that is 0.000000000000000000000000000000000001 s old).

The period of cosmic inflation, Guth argued, only lasted until the Universe was about 10^{-33} or 10^{-32} s old, which of course is an incredibly brief period. But, actually, at the end of the inflation period, the Universe was about one thousand times older than it had been at the beginning of the inflation period. More importantly though, it was much bigger. At the start of the inflation period Guth argued that the Universe was smaller than an atomic proton, so let us say about 10^{-15} m in diameter. By the end, it was about the size of a marble, so let us say about 2 cm in diameter. That is an increase in size of about a factor of 10^{13} (10,000,000,000 or 10 trillion). To put this into some kind of context, this is about the same as the ratio of the diameter of a marble to the distance from the Earth to the Sun! Guth argued that after this brief period of hyper-expansion, the Universe settled down to its normal rate of expansion [20].

4.4.4.1 How Inflation Solves the Horizon Problem

The horizon problem is solved by inflation because the very rapid expansion which inflation proposes would allow parts of the Universe which are now too far apart to have ever communicated with each other to have been close enough together before inflation. Going back to my analogy with the room being heated, it is as if the room started off being really small, let's say just 0.05 m by 0.05 m by 0.01 m (I know one cannot fit a heater into such a small room, but this is an analogy!). Let us suppose it started off this tiny, an suddenly after 1 s it rapidly expanded to being 5 m by 5 m by 2 m in size in a fraction of a second. Before the rapid expansion started the tiny room had enough time for the heat to spread into all the corners, and so even when the room has grown to its larger size the corners would all have the same temperature. When we measure the temperature in the corners 30 s after the heater was turned on we find that all the corners have the same temperature, and it is the rapid expansion enabled this temperature uniformity.

4.4.4.2 How Inflation Solves the Flatness Problem

The flatness problem is solved by inflation by drawing the analogy between the geometry of the Universe and a curved surface. If a curved surface is large enough, then on a local scale it is always going to look flat. An easy analogy to understand this is the surface of our Earth. We all know it is spherical, but on a local scale it appears flat. Your dining room floor is flat, your local football field is (hopefully!) flat. And yet we know that both are on a curved surface.

If the Universe underwent a period of cosmic inflation, then we are seeing only a tiny part of it, what we call our *'observable Universe'*. No matter what the geometry of the Universe as a whole, inflation argues that the part we can see is going to appear flat. If you look at Fig. 4.6 again, and concentrate on the small grid pattern on each of the three geometrical surfaces, if that grid pattern were vanishingly small then each square in the grid pattern would be flat, no matter what the geometry of the overall surface.

4.4.4.3 How Inflation Solves the Magnetic Monopole Problem

The magnetic monopole problem is solved by cosmic inflation in a very elegant way, and of course it was thinking about this problem that had got Guth thinking about inflation in the first place. The idea he proposed is that magnetic monopoles were created in large quantities before the period of cosmic inflation. They should still exist today, but because the Universe expanded so rapidly during cosmic inflation, their number density (how many there are per unit volume) is so tiny that we haven't found any in the part of the Universe which we are able to observe.

4.5 Studying the CMB from Space

By the late 1960s theoreticians were realising that the structure of galaxies and clusters of galaxies which we see in the Universe must have existed at the time of the CMBR. Although the CMBR was produced at a time long before the formation of any stars or galaxies, the Universe must have contained the density variations which were the seeds of the structure we see today, and these density fluctuations should show up as temperature variations in the CMBR. English theoretical astrophysicist Joe Silk wrote one of the more important papers on this topic in Nature magazine in 1967, entitled *"Primordial fluctuations in the CMBR"* [21]. As of 1977, the only variations which had shown up in the isotropy of the CMBR was the dipole variation that Smoot and colleagues had found [5], which was attributed to our Milky Way moving towards Leo at a speed of some 600 km/s. If there were any more subtle variations, they had eluded detection.

Experimental cosmologists realised in the mid 1970s that they were reaching the limit of what could be done from either the ground or high in the stratosphere from U2 aeroplanes or balloons. To achieve the sensitivity necessary to find the small variations in the CMBR that theory was saying existed, they would have to put an experiment into space. In 1974 NASA made a call for space-based astronomy experiments, and Smoot and his colleagues at the LBNL submitted a proposal to launch a satellite to look for temperature variations in the CMBR. Two other teams also submitted similar proposals, a team including Samuel Gulkis and Mike Janssen at NASA's Jet Propulsion Laboratory in Pasadena, and a team led by Mike Hauser and colleagues at NASA's Goddard Space Flight Center (GSFC) in Maryland. One of the members of Hauser's team was John Mather (Fig. 4.8), who in 1974 was working as a Postdoctoral Fellow at NASA's Goddard Institute for Space Studies in New York City adjacent to Columbia University, having just got his Ph.D. in physics from UC Berkeley.

Mather was born in 1946 in Roanoke, Virginia. When Mather was born his father, Robert Mather, had just started a faculty position at what would become Virginia Tech. Robert Mather had obtained his Ph.D. from the University of Wisconsin and was embarking on a research career in animal husbandry and statistics when his son entered the World. Mather's mother, Martha, was a high school French teacher. Mather showed an early interest in science, he recalled later in an interview how at the age of six he started writing out the integers on a sheet of paper, and realised that even by the time he had filled the page he had not reached a very high number. He said that this gave him an early appreciation of infinity!

As is true of many people who go on to a career in astronomy, Mather had a telescope as a child, and his father also bought him the book *"Astronomy Made Simple"* [22], which he devoured. Between his penultimate and final year of high school, Mather attended a summer physics programme at Cornell University, and then during his senior high school year he chose Swarthmore College on the outskirts of Philadelphia for his undergraduate studies, entering in 1964. In 1993, nearly 30 years later, I was lucky enough to spend a year at Swarthmore as an

Fig. 4.8 John Mather, principal investigator on COBE's FIRAS instrument, in his office at Goddard space flight center (image credit: NASA)

assistant professor, and I can vouch for the very high calibre of the students who go to this highly selective small liberal arts college. Mather was in the honours programme at Swarthmore, and graduated in 1968 with highest honours.

He had decided to go to nearby Princeton for his graduate studies, but during the spring of his final year at Swarthmore a childhood friend Ted Chang sent him a photograph of Chang sitting in a T-shirt next to the University of California Berkeley campus fountain in January. After graduating from Swarthmore, Mather got a summer job working at the LBL, which gave him a chance to see Berkeley first hand. After spending his summer there, he decided to switch his graduate studies to Berkeley, and he started there in the autumn of 1968. He went there intending to specialise in particle physics, but in 1970 he got chatting to a young faculty member, Paul Richards, who was starting to work on a project to measure the spectrum of the CMBR with Charles Townes and postdoctoral researcher Michael Werner. The purpose of the experiment was to see how close to a blackbody was the spectrum of the CMBR.

4.5 Studying the CMB from Space

Mather became fascinated by the idea of this project, and so started work on building the instrumentation to measure the spectrum of the CMBR. They took their experiment to the summit of White Mountain in eastern California, which at an altitude of 4,344 m (14,252 ft) is above most of the Earth's atmosphere. However, even at this altitude, measurements of the CMBR are affected by emissions from the atmosphere, so the experiment was only able to set some limits on the CMBR's intensity at different wavelengths [23]. The Berkeley team decided they needed to try a balloon-borne experiment to improve their sensitivity, so Mather and fellow graduate student David Woody set about designing, building and testing the new experiment. Unfortunately the experiment failed, so Mather wrote his Ph.D. thesis on the results from their experiment on White Mountain and on the design of the balloon-borne experiment [24]. He was offered a post doctoral position at the Goddard Institute for Space Studies in New York City, which is where he was working when NASA made its call for proposals in 1974, and this call for proposals got Mather thinking about doing the experiment to measure the CMBR's spectrum from space.

In 1976 NASA decided to get the three groups who had proposed to do space-based CMBR experiments to work together. They appointed a committee made up of Gulkis, Hauser, Mather, Rainer Weiss, Dave Wilkinson and Smoot to evaluate and study ideas for a future mission. It was at this point that Hauser persuaded Mather to leave New York City and move to GSFC in Maryland. The mission the committee recommended was to have a satellite with three experiments on board. Hauser would head up an experiment called the *'Diffuse Infrared Background Explorer'* (DIRBE), the purpose of which was to map the dust emission of the Milky Way galaxy from $1.25\,\mu$ in the near infrared all the way out to $240\,\mu$ in the far infrared. Mather would head up an experiment called the *'Far Infrared Absolute Spectrometer'* (FIRAS), which would measure the complete spectrum of the CMBR, and finally Smoot would head up an experiment called the *'Differential Microwave Radiometer'* (DMR), which would look for anisotropies (variations) in the temperature of the CMBR. The name given to the whole mission was the *'Cosmic Background Explorer'* (COBE). By 1980 the team had convinced NASA of their recommendation, and were told that COBE would be launched on the Space Shuttle in 1988.

On the 28th of January 1986 I was working in London (as a trainee accountant, but that's another story!), and came home after a long day's work to watch the evening news at 9 pm. I had not heard any news since the morning, and had motorcycled back from work in the dark, so had not picked up the evening newspaper as I sometimes did when I was using the London Underground. As I sat down to watch the news, the opening video footage left me in a state of shock. The Space Shuttle Challenger had exploded a few minutes into its launch earlier that day, with all seven people on board being killed in the explosion. Not only was the accident a tragedy for the families who lost their loved ones but, until the cause of the accident could be determined, all planned space shuttle launches were postponed.

Then, a few months later, on the 19th of April, a Titan rocket exploded on take off, followed a few weeks later on the 3rd of May by a Delta rocket also exploding on take off. Although the Titan and Delta rocket explosions did not involve any loss of life, three successive disasters left NASA's launch capabilities in jeopardy. As a senior executive at McDonnel Douglas, who built the Delta rocket, told the United Press Association, "The aerospace industry was in chaos" [25]. Not only were all space shuttle launches indefinitely postponed, but NASA now suspended all rocket launches too.

Worried that this might spell the end of their satellite, the COBE scientists considered asking the French to launch their satellite using an Ariane rocket from their facility in French Guyana. When NASA heard of this they decided to give the go-ahead for COBE to be launched on the last of the small Delta rockets, rather than face the embarrassment of having a major American science satellite be launched by a French rocket. On the 1st of October 1986 NASA officially told the COBE team that they would launch in early 1989 on a Delta rocket, but this rocket could only lift half the payload of the space shuttle. It would mean that the COBE team would have to completely redesign the three instruments to reduce their weight and volume to fit the new launch vehicle.

Three years didn't give the COBE team much time to redesign their instruments and satellite. All the work would be done at Goddard, with each instrument being redesigned, developed, tested and built in three adjoining buildings. Smoot remained the Principal Investigator (PI) of the DMR, with Chuck Bennett working as his deputy. Mather remained the PI of FIRAS, with Rick Shafer as his deputy, and finally Hauser remained PI of DIRBE, with Tom Kelsall as his deputy. The COBE team did not underestimate the huge challenge facing them, but Goddard executives were also fully aware of the challenge. In the Goddard Engineering Newsletter it was stated

> The transformation of COBE from an STS [space transportation system, that is a space shuttle launch] to a Delta launch is probably one of the greatest engineering challenges ever undertaken by the GSFC. [26]

4.6 Waiting for COBE

As anyone who has worked on space-based projects knows, the time between beginning the project and the results coming in is usually well over 10 years. This is mainly because the instruments used in any space-based mission have to be very thoroughly tested, as in most cases there is no possibility of fixing anything once it is in space should it go wrong. When a satellite is launched on a rocket, the vibrations are sufficiently strong to destroy most instruments built for ground-based work, and so space-based instruments need to be built to withstand much higher levels of mistreatment. Because space-based instruments cannot normally be fixed once they are launched, space-based experiments tend to use more tried and tested technology, as it is usually more reliable.

4.6 Waiting for COBE

Most astronomers accept that the extra sensitivity which can be gained by going into space is worth both the additional time involved and the use of older technology. By the time the rocket carrying a space-based experiment launches, the instrumentation on it is typically over a decade old. It is true that most current technology is nearly always more advanced both in terms of sensitivity and, often, the number of detectors which can be included in a detector array, but it is felt that this is usually more than offset by the improvements in sensitivity possible from the benign environment of space.

In 1980 the CMBR world was rocked by reports of anisotropies in the CMBR by two teams. The first team was based in Italy, led by Francesco Melchiorri, who at the time was at the University of Florence; the second was a US based team led by Wilkinson of Princeton. Both teams reported a quadrupole in the CMBR, which means two hotter than average spots and two colder than average spots. The dipole discovered by Smoot and his team had been explained by our Galaxy's motion towards Leo, but the additional warm and cold spots reported separately by Melchiorri et al. [27] and Wilkinson et al. [28] could not be wholly attributed to this motion. If it was real it was possibly the first signs of the long-sought after temperature variations which would have been the seeds of cosmic structure.

Smoot and his colleagues decided they needed to check the results. By this time they had developed more sensitive differential microwave radiometers (DMRs), cooled by liquid helium and operating at only 4 K. By cooling the detectors to liquid helium temperatures the noise (mainly due to the thermal motion of the components in the detector) could be reduced to much lower levels, this dramatically increases their sensitivity. They first tried to detect the quadrupole using the U2 aeroplane, but did not detect it. They then decided they needed to further improve their sensitivity by getting even higher in the atmosphere, which would reduce even further the background from atmospheric emissions. The only way to do this was to do a balloon-borne experiment.

Smoot had previous bad experiences with balloon-borne experiments during his time working on searches for anti-matter in the early and mid 1970s, but decided that checking for a possible quadrupole in the CMBR was important enough that he was willing to return to the frustrations of them. In April 1982 his team flew an experiment from the US National Scientific Ballooning Facility in Palestine, Texas. Although they clearly detected the already known dipole, which was a good indication that the experiment had worked, they failed to detect any quadrupole signal. To complete the experiment, however, they needed to also fly an experiment to observe the southern sky. In November of the same year they flew their experiment from Institutio de Pesquisas Espaciais (INPE) based in Cachoeira Paulista in Brazil, which is about 225 km (150 miles) west of Rio de Janeiro and slightly inland.

Unfortunately, after a successful night of making observations, the gondola carrying the experiment failed to drop from the balloon as planned. When the INPE crew member pressed the button to trigger the release of the gondola, nothing happened. The balloon was also fitted with a backup system which would trigger the release mechanism automatically if the manual release failed, but this too did not work. Smoot and his colleagues sat helplessly in the control centre at IPNE as they watched their valuable data disappear above the Brazilian jungle. Despite attempts by the Brazilian air force to find the balloon and its precious cargo, all attempts failed and the team had to accept defeat.

Smoot and his LBL colleagues had to satisfy themselves with publishing the results from an incomplete experiment. Although they were able to report that they did not detect the quadrupole as Melchiorri's team and Wilkinson's team had reported, they had to add the caveat that they had no observations of the southern sky to strengthen their claims. Nearly 18 months later, in January 1984, Smoot received a telephone call to say that the Brazilian authorities had found the balloon. It had been found deep in the jungle near a small village by the name of Tapira, about 500 km west of Rio, and 225 km west of where it had been launched in Cachoeira Paulista. When the team recovered the tape carrying the data, and got it back to Berkeley, they found it to be 98 % usable, despite having green fungus growing on it from some 18 months in the Brazilian jungle! When the data were analysed, they showed the same thing that Smoot and his colleagues had found for the observations of the northern hemisphere, there were no signs of any quadrupole.

4.7 A Perfect Blackbody

Finally in late 1989, after many years of effort, and over 1,000 people involved, COBE was ready for launch. The project had cost about $160 million, and would be the first space-based mission to study in detail the CMBR which Penzias and Wilson had accidentally discovered in the mid 1960s. And, in a wonderful twist of fate, it would be launched on the last ever Delta rocket that McDonnel Douglas were to build, with the Echo 1 satellite being launched on the first ever Delta rocket. The Holmdale horn which had discovered the CMBR had been specifically built to send and receive signals from the Echo 1 satellite.

The satellite was put together at Goddard and flown from Andrews Air Force Base to Vanderberg Air Force Base on the California coast. At about 6:30 am Pacific Standard Time on the 18th of November 1989, COBE soared up into space on top of its Delta rocket, and was put into what is called a 'polar orbit', about 900 km (550 miles) above the Earth's surface. A polar orbit, as the name implies, is one which takes a satellite over the Earth's north and south poles. It is the preferred orbit for spy satellites, as on each orbit they will pass over a different part of the Earth. In COBE's case, of course, it was looking away from the Earth at the oldest light in the Universe, and the polar orbit it was in meant it took just over 100 min to make one complete orbit of the Earth. As the satellite orbited the Earth it also rotated, roughly once per minute. This was partly to keep it stable, and partly to ensure even heating of the satellite by the Sun, but it also meant that each of the three instruments could efficiently map the whole sky. It would take 6 months for each instrument to map the whole sky, but they would continue mapping the sky several times to increase the sensitivity of their results.

4.7 A Perfect Blackbody

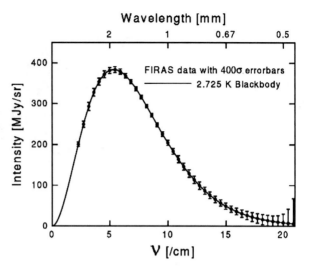

Fig. 4.9 The spectrum produced by COBE's FIRAS instrument. The data show a perfect blackbody curve with a temperature of 2.725 K. The error bars shown are 400σ. Usually error bars of 3 or 5σ are sufficient, but these would be too small to see in this case (image credit: NASA/COBE/FIRAS)

The COBE team had decided to give an announcement of preliminary results at the January 1990 meeting of the American Astronomical Society (AAS), which was being held in Washington, DC [29–31] The team had also agreed that, before they made any announcement at the meeting, they would submit papers to the Astrophysical Journal (ApJ). The meeting was barely 2 months after the launch of COBE, and so tight was the deadline that Smoot and Mather dropped off their papers to the editors of ApJ on their way to give their talk at the AAS meeting; their talk was scheduled on the Saturday, the last day of the week long meeting. As they walked into the room where they would speak, Smoot and Mather found a room crammed with over one thousand people.

Mather took to the podium, and after a short introduction he showed a theoretical graph of what a blackbody spectrum should look like. Superimposed on the theoretical curve were sixty seven measurements made by the FIRAS instrument, all at different wavelengths (Fig. 4.9). The data fitted the curve perfectly, in fact it is the most perfect blackbody spectrum scientists have ever measured. FIRAS had measured the spectrum of the CMBR to be a blackbody with a temperature of 2.735 K. The audience rose and burst into a round of applause, this was a stunning vindication that the CMBR was indeed the afterglow of the creation of the Universe, and the result was a strong confirmation that the big bang theory was indeed correct.

After the applause had died down, Smoot took to the podium and showed the preliminary results from the DMR experiment, the one that the COBE team hoped 1 day would show the long-sought after anisotropies. After only 2 months of gathering data, it was still too early to see such details in the map, if they existed at all. However, the map that Smoot did show the audience was still an important one, even if it was less spectacular than the stunning blackbody curve his colleague Mather had just shown. Smoot was able to show that COBE was seeing the dipole that he and his colleagues had announced some 13 years before in April 1977, but the COBE map of

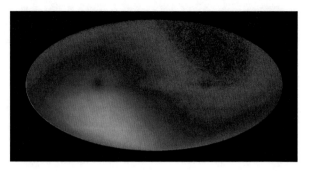

Fig. 4.10 The dipole moment as seen by the DMR experiment (image credit: NASA/COBE/DMR)

the dipole (Fig. 4.10) was of a better quality than any previously seen. Importantly, it showed that the DMR instrument was working correctly, and the audience knew that it would take many more months of observations to reach the sensitivity to be able to see any anisotropies, if they existed at all. By a peculiar twist of fate, the person chairing this session of the AAS meeting was Geoffrey Burbidge, who along with Fred Hoyle and a few others had been one of the architects of showing that most of the elements were formed in stars, and like Hoyle had been a life-long advocate of the competing steady state theory. As Smoot got down from the podium he heard Burbidge let out an audible sigh, it was clear to the supporters of the steady state theory that COBE had started producing results which was going to make their theory increasingly untenable.

There still remained several steps in the quest to show that the CMBR was indeed not only the oldest light we can observe, but also our first snapshot of the infant Universe with the imprint of later structure already in it. The uniformity of the CMBR had been shown by Partridge and Wilkinson in 1967, the dipole had first been seen by Smoot and his collaborators in 1976 using their DMR in the U2 spy-plane. The quadrupole, which had been erroneously reported by Melchiorri's team and Wilkinson's team but refuted by Smoot and his team's more sensitive measurements, was still to be confirmed, and of course the final step was to find the anisotropies in the CMBR which would indicate the first signs of the density variations which would later grow to become the stars, galaxies, clusters and superclusters that we observed.

4.8 The "Scientific Discovery of the Century"

By early 1991, just over 12 months after it had started taking data, the DMR instrument was beginning to see the quadrupole in the CMBR. This was the first sign that there was indeed structure beyond the structure due to our Galaxy's non Hubble-flow motion towards the Great Attractor. The DMR carried on making observations, and was slowly building up enough data that the actual anisotropies, the finer-scale structure, should start to become visible. By the summer of 1991,

4.8 The "Scientific Discovery of the Century"

with no announcement of any anisotropies from the COBE team, the scientific press was beginning to speculate that they had not found any. If this were the case, the big bang theory would be in crisis. Despite the stunning confirmation that the CMBR had a blackbody spectrum, without the anisotropies how could adherents of the big bang theory explain the cosmic structure we see in our Universe?

Slowly, very slowly, in the autumn of 1991, the DMR team started to see the first evidence for the elusive anisotropies. In October 1991, Ned Wright of UCLA, who was a member of the DMR team, wanted to publish this preliminary data, more than anything to stop the rampant speculation in the astronomical community which was suggesting that they had found nothing. However, Smoot and Wilkinson argued against this, wary of the embarrassment an incorrect announcement would make, they wanted to be more certain that the emerging results were real.

One of the questions which plagued Smoot was the possibility that the structure they were seeing was due to contamination from our Galaxy. In order to produce their map of the CMBR they had to remove the signal from our Milky Way, but unfortunately no-one had produced a radio map of the Milky Way at the wavelengths that the DMR was observing. The DMR had six receivers working in pairs, and observed at frequencies of 31.5, 53 and 90 GHz, which correspond to wavelengths of 0.0095, 0.0057 and 0.0033 m, all wavelengths that it was nigh-on impossible to observe from the ground as the emission from the atmosphere was so strong. This is why there were no maps of the Milky Way at these wavelengths.

In the absence of any maps, the DMR team had to interpolate from already existing maps at other wavelengths, and make assumptions about the strength of the Galactic emission at the wavelengths the DMR instrument was using. Although this was a reasonable thing to do, it was not foolproof. Any errors in the interpolation of the data could lead to a false signal, which the DMR team could be misled into thinking was indications of anisotropies, when in fact they were contamination from our Galaxy which had been incorrectly removed. It was imperative to get this right, they were looking for tiny variations in the uniform temperature of the CMBR, thousandths of a Kelvin variations in the 2.735 K temperature.

There was only one thing for it, Smoot and his team would have to observe the Milky Way galaxy at the wavelengths the DMR was using, and yet the atmosphere's emission precluded them. Nearly. There was one place on the Earth where the atmosphere is so cold that it emits significantly less in this part of the spectrum that anywhere else on Earth, and that is Antarctica. For reasons I will discuss more detail in Chap. 5, the Antarctic continent provides the closest thing we have here on Earth to conditions in space for many types of astronomical observations, and it is for these reasons that the DMR team found themselves heading to the South Pole in November 1991.

Before leaving for the South Pole, they had in Berkeley designed and built a 10-m diameter radio dish in segments, and this was shipped down to the South Pole along with several members of the team. In late November and early December 1991 the radio telescope observed the Milky Way at five different wavelengths, but unfortunately three of the channels were contaminated by what appeared to be satellites passing periodically overhead, and so only two of the five channels were

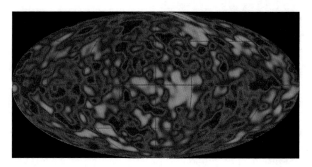

Fig. 4.11 COBE's map of the temperature fluctuations in the cosmic microwave background. The fluctuations are only one part in 100,000 about the average temperature of 2.725 K. Hotter areas are *red*, colder areas are *blue*. These fluctuations are the seeds of the structure we see in the Universe today (image credit: NASA/COBE/DMR)

usable. They got three weeks of data, but it was sufficient to be able to produce the accurate radio map of the Milky Way that the COBE DMR team needed to do the subtraction of the Galactic emission properly. When they did, the anisotropies were still there. They had found the *"Holy Grail of cosmology"*, or at least the Holy Grail of cosmology as it was in the 1970s, 1980s and 1990s (Fig. 4.11).

The announcement of their result that COBE had indeed found the beginnings of the present-day structure in the Universe was to be made on Thursday the 23rd of April 1992 at a meeting of the American Physical Society in Washington D.C. [32] The meeting 15 years to the month that Smoot had first shocked the cosmology community when he announced the dipole in the CMBR from their U2 DMR experiment. Although significant, that announcement in 1977 would pale into insignificance compared to the announcement he would make on finding the anisotropies.

I remember the day very well, as we in the Physics and Astronomy Department at Cardiff University had been busy preparing for a visit of the then Astronomer Royal, Prof. Arnold Wolfendale, to our department on that day. It was a big deal to have the Astronomer Royal visit a department, and we had spent weeks preparing the most impressive demonstrations we had on our research to impress him. The day started off normally, with various research groups in the department showing Prof. Wolfendale the results of their latest observations and theories. He was slowly working his way around the numerous groups, and my group was scheduled to show him what we had been working on early in the afternoon.

After a break for lunch, I got myself ready to show him the work I had been doing, and ran through the images I was going to show him for the last time, practicing what I was going to say as each image came up on the screen. He never made it to see me, less than half an hour into this afternoon session the news broke of the stunning finding by the COBE team, they had finally found the anisotropies in the oldest light in the Universe. Within minutes of their announcement in Washington, DC the British newspapers and radio and television reporters had tracked down

Prof. Wolfendale to Cardiff, and he spent the rest of the afternoon doing interviews to explain one of the most important discoveries in cosmology to the wider public.

Although I was disappointed I didn't get to show him my carefully prepared presentation on my research, I too was caught up in the excitement of the day. We all gathered around a television set that someone quickly set up in our communal area, and we watched as the BBC reported on the story. It wasn't long before quotes from famous scientists started appearing on the television, and one of the more memorable ones was by Stephen Hawking, who had gained fame for his book '*A Brief History of Time*'. Hawking referred to the COBE discovery as

> the scientific discovery of the century, if not of all time

I remember at the time laughing at this, how could the discovery of variations in the CMBR be more important that the discovery of the CMBR itself? Of course it couldn't be, but it made for a wonderful soundbite, and it was indicative of just what an important results this was for the world of cosmology, and for supporters of the big bang theory. Finally, we could point to the oldest light and say that indeed we had seen the seeds of the structure which today we see all around us in the vast Cosmos.

References

1. Sciama, D.W.: Peculiar velocity of the Sun and the cosmic microwave background. Phys. Rev. Lett. **18**, 1065–1067 (1967)
2. Partridge, R.B., Wilkinson, D.T.: Isotropy and homogeneity of the Universe from measurements of the cosmic microwave background. Phys. Rev. Lett. **18**, 557–559 (1967)
3. Peebles, P.J.E.: Physical Cosmology. Princeton University Press, Princeton (1971)
4. Smoot, G.F., Davidson, K.: Wrinkles in Time. William Morrow, New York (1994)
5. Gorenstein, M.V., Smoot, G.F., Muller, R.A.: Anisotropy of the cosmic microwave background radiation. Bull. Astron. Soc. **9**, 431 (1977)
6. Zwicky, F.: Die Rotverschiebung von extragalaktischen Nebeln ("The redshift of extragalactic nebulae"). Helv. Phys. Acta **6**, 110–127 (1933)
7. Zwicky, F.: On the masses of nebulae and of clusters of nebulae. Astrophys. J. **86**, 217–246 (1937)
8. Ostriker, J.P., Peebles, P.J.E.: A numerical study of the stability of flattened galaxies: or, can cold galaxies survive? Astrophys. J. **186**, 467–480 (1973)
9. Rubin, V.C.: Fluctuations in the space distribution of the Galaxies. Ph.D. thesis, Georgetown University (1954)
10. de Vaucouleurs, G.: The distribution of bright galaxies and the local supergalaxy. Vistas Astron. **2**, 1584–1606 (1956)
11. Rubin, V.C., Thonnard, N., Ford, W.K.: Extended rotation curves of high-luminosity spiral galaxies. IV—Systematic dynamical properties, Sa through Sc. Astrophys. J. **225**, L107–L111 (1978)
12. Rubin, V.C., Ford, W.K., Thonnard, N.: Rotational properties of 21 SC galaxies with a large range of luminosities and radii, from NGC 4605 (R − 4 kpc) to UGC 2885 (R = 122 kpc). Astrophys. J. **238**, 471–487 (1980)
13. Bosma, A.: The distribution and kinematics of neutral hydrogen in spiral galaxies of various morphological types. Ph.D. thesis, University of Groningen (1978)

14. van der Kruit, P.C., Bosma, A.: The rotation curves and orientation parameters of the spiral galaxies NGC 2715, 5033 and 5055. Astron. Astrophys. Suppl. Ser. **34**, 259–266 (1978)
15. Davis, M., Huchra, J., Latham, D.W., Tonry, J.: A survey of galaxy redshifts. II—The large scale space distribution. Astrophys. J. **253**, 423–445 (1982)
16. Huchra, J., Davis, M., Latham, D., Tonry, J.: A survey of galaxy redshifts. IV—The data. Astrophys. J. Suppl. Ser. **52**, 89–119 (1983)
17. Huchra, J., Geller, M., de Lapparent, V., Corwin, H., Jr.: The CfA redshift survey—data for the NGP +30 zone. Astrophys. J. Suppl. **72**, 433–470 (1990)
18. Huchra, J.P., Vogeley, M.S., Geller, M.J.: The CFA redshift survey: data for the south galactic CAP. Astrophys. J. Suppl. **121**, 287–368 (1999)
19. Geller, M.J., Huchra, J.P.: Mapping the universe. Science **246**, 897–903 (1989)
20. Guth, A.H.: Inflationary universe: a possible solution to the horizon and flatness problems. Phys. Rev. D **23**, 347–356 (1981)
21. Silk, J.: Fluctuations in the primordial fireball. Nature **215**, 1155–1156 (1967)
22. Degani, M.H.: Astronomy Made Simple. Made Simple Books, New York (1955)
23. Mather, J.C., Werner, M.W., Richards, P.L.: A search for spectral features in the submillimeter background radiation. Astrophys. J. **170**, L59–65 (1971)
24. Mather, J.C.: Far infrared spectrometry of the cosmic background radiation. Ph.D. thesis, University of California, Berkeley (1974)
25. Smoot, G.F., Davidson, K.: Wrinkles in Time, p. 220. William Morrow, New York (1994)
26. Smoot, G.F., Davidson, K.: Wrinkles in Time, p. 224. William Morrow, New York (1994)
27. Melchiorri, F., Ceccarelli, C., Pietranera, L., Melchiorri, B.O.: Fluctuations in the microwave background at intermediate angular scales. Astrophys. J. **250**, L1–L4 (1981)
28. Boughn, S.P., Cheng, E.S., Wilkinson, D.T.: Dipole and quadrupole anisotropy of the 2.7 K radiation. Astrophys. J. **243**, L113–L117 (1981)
29. Bennett, C., Smoot, G., Kogut, A., Jackson, P., Rokke, L., Backus, C., Galuk, K., Huang, Q., Keegstra, P., Aymon, J., De Amici, G., Tenorio, L., Gulkis, S., Janssen, M., Lubin, P.: Results from the COBE DMR instrument at "Six Months" after launch. Bull. Am. Astron. Soc. **22**, 874 (1990)
30. Wright, E.L., Cheng, E.S., Dwek, E., Bennett, C.L., Boggess, N.W., Mather, J.C., Shafer, R. A., Hauser, M.G., Kelsall, T., Moseley, S.H., Jr., Silverberg, R.F., Smoot, G.F., Eplee, R. E., Issacman, R.B., Meyer, S.S., Weiss, R., Gulkis, S.G., Janssen, M., Lubin, P.M., Murdock, T.L., Wilkinson, D.T.: Preliminary millimeter and sub–millimeter COBE observations of the milky way galaxy with a 7° beam. Bull. Am. Astron. Soc. **22**, 874 (1990)
31. Cheng, E.S., Bennett, C.L., Boggess, N.W., Dwek, E., Gulkis, S., Hauser, M.G., Janssen, M., Kelsall, T., Lubin, P.M., Mather, J.C., Meyer, S.S., Moseley, SH., Jr., Murdock, TL., Shafer, R.A., Silverberg, R.F., Smoot, G.F., Weiss, R., Wilkinson, D.T., Wright, E.L.: Status of the COBE satellite. Bull. Astron. Soc. **22**, 876–877 (1990)
32. Smoot, G.F.: COBE DMR observations of early universe physics (Proceedings of the XXVI International Conference on High Energy Physics, vol. II). AIP Conf. Proc. **272**, 1591–1601 (1992)

Chapter 5
To the Ends of the Earth

The quantum scale fluctuations which produced the anisotropies in the cosmic microwave background also set up acoustic oscillations in the gas of the early Universe, and these acoustic oscillations left their imprint on the cosmic microwave background. After COBE, studying this finer level detail in the anisotropies became the focus of experimental cosmology. Several experimentalists wanted to go back into space with a more capable satellite than COBE, and so the "*Microwave Anisotropy Probe*" (MAP) was conceived. During its development, ground-based experiments from the South Pole started probing details of the CMB, and the Center for Astrophysical Research in Antarctica was established to coordinate this work. Other ground-based work was being done in the high desert in northern Chile, and it was there that the first hints were seen of the so-called "*first acoustic peak*" in the anisotropies that theoreticians had predicted. A balloon-borne experiment BOOMERanG launched from the Antarctic content saw strong evidence in 1998 that the Universe is indeed flat. In the same year, the astronomical community were shocked to learn that the Universe's expansion is speeding up, and the idea of "dark energy" was introduced to explain it. In 2001 the DASI telescope at the South Pole found the first evidence of polarisation in the CMB, another important vindication of a prediction of the big bang theory.

5.1 Primordial Sound Waves

The DIRBE instrument on the COBE satellite which observed the anisotropies in the CMB had a beam size of 7°. This meant that it was unable to see any variations in the CMB on angular scales any smaller than this. Inflation theory had predicted that the number of fluctuations at different angular scales would be roughly constant, and COBE had shown that this appeared to be the case for angular scales of 7° and greater. But the filaments and voids and superclusters and clusters of galaxies which astronomers see in the Universe today meant that most of the structure we see in the Universe today existed in the CMB at angular scales much smaller than COBE's rather crude resolution. Following on from COBE's success, it was imperative that these smaller angular scales be investigated.

Before decoupling when the CMB radiation was produced, the Universe was a plasma. This trapped the radiation very effectively, leading to a tightly coupled system where the photons, electrons and protons were constantly interacting with each other, behaving like a single gas. It was realised in the late 1960s by Jim Peebles and his PhD student Yu Jer that this would mean that the very early Universe would contain sound waves, or as physicists tend to call them, *"acoustic waves"*. They published this idea in a paper entitled *"Primeval Adiabatic Perturbation in an Expanding Universe"* [1] which appeared in 1970 in the *Astrophysical Journal*. About the same time the effect was also realised by the Russian theoretical cosmologists Rashid Sunyaev and Yakov Zel'dovich, who are better known for the Sunyaev-Zel'dovich effect that I discuss in Sect. 7.1. They published their theory in the same year in a paper entitled *"Small-scale Fluctuations of the Relic Radiation"* [2], which appeared in the journal *Astrophysics and Space Science*. Because of this predicted behaviour, any small disturbances in the primordial gas would propagate through the gas as acoustic waves, and the physics of acoustic waves is very well understood.

It is the production of acoustic waves in, for example, the air in a room which transmits sound from its source to our ears. The vibrations where the sound is being produced set up acoustic waves in the air, and these transfer the vibrations from the source of the sound to our ears. Many of us remember seeing the demonstration of a bell ringing inside a glass jar. As the air is sucked out of the glass jar the sound of the bell gets fainter and fainter, and it eventually becomes silent when the air is fully evacuated, even though the bell is clearly still ringing. With no air inside of the jar, the vibrations have no way to travel from the bell to the glass of the jar, and hence we hear no sound. But, when the glass jar has air in it, we can hear it. The vibrations in the bell set up acoustic waves in the air in the jar, and these waves in turn cause the glass itself to vibrate. The air between the outer surface of the glass and our ears transmits the sound to our ears via acoustic waves in the air in the room, and we hear the sound of the bell ringing.

Unlike water waves on the surface of a lake, which are an example of *transverse waves*, acoustic waves are what we call *longitudinal waves*. Acoustic waves transfer their energy by alternatively compressing and rarifying the air, so any given patch of air in the room alternates between being compressed and being rarefied. The air next to our ear drum is, therefore, also alternatively compressed and rarified, and this causes it to vibrate, mimicking the vibrations of the source of the sound. How would acoustic waves in the early Universe affect the gas? The initial fluctuations

5.1 Primordial Sound Waves

in the Universe are believed to be due to quantum mechanics, what we call *quantum fluctuations*, and it is these initial quantum fluctuations would have produced acoustic waves in the primordial gas of the early Universe. The compression and rarefaction of these acoustic waves would have heated and cooled the gas respectively, hot spots being created in areas of compression and cool spots in areas of rarefaction. In this way, fluctuations in the early Universe resulted in a shifting pattern of temperature fluctuations in the expanding gas.

As we saw in Sect. 3.5, the blackbody radiation produced by objects looks exactly the same no matter whether the object is a star or a canon ball or a hot car engine. Through studying the blackbody radiation there is no way to determine anything about the composition of the emitting object, just its temperature. The converse is true when we study the acoustic waves in an object. For example, even our ears can easily distinguish between the sound of the same piece of music being played on a piano or on a harpsichord. Even though the two instruments are playing the same notes, they sound different. At a more subtle level, people with a good musical ear can distinguish between the sound made by a Stradivarius violin and other less celebrated ones. Different guitars will produce a different sound when they play the same music, a Gibson guitar sounds different to a Fender Stratocaster. Why is this?

The answer is that the different instruments are vibrating in different ways. When a musical instrument plays a particular note, it also produces what we call *harmonics*, or *overtones*. The strength of these harmonics compared to the main note, which is called the *fundamental mode*, will determine how the instrument sounds. Because the nature of the instrument determines its sound, we can work backwards from the sound produced by a musical instrument to determine its nature—we can deconstruct its composition and its shape. It has been suggested that the supposedly superior sound of a Stradivarius violin is due to the way the wood was treated with either a varnish or lacquer when made. A detailed analysis of the fundamental mode and harmonics for a Stradivarius would show that they are different to those of an inferior violin.

The Sun is, like the early Universe, a plasma, and in "*helioseismology*" physicists study the pattern of vibrations at the Sun's surface to learn about its internal structure. The details of the vibrations at the surface are dependent on the temperature and density profile of the gas within the Sun, and on which parts of the interior are convective and which parts are radiative. By analogy, people started thinking that maybe the acoustic modes of the early Universe could be used to tell us more about its composition. The wavelength of the early Universe's fundamental mode would show up on the pattern of the CMB as the largest temperature fluctuations. It was realised that the angular scale of the fundamental node, which became known as the "*first acoustic peak*", would depend on the geometry of the Universe. Furthermore, the angular scale and strength of the higher harmonics would also be influenced by the composition of the Universe. A Universe with a lot of normal baryonic matter would have different higher order acoustic modes to a Universe which whose matter content was dominated by dark matter.

So, whilst experimentalists set about devising and building the next generation of instruments to probe this smaller scale structure of the CMB, theoreticians worked on refining their theories to make predictions of what the CMB should look like at these different angular scales. They produced different models to predict the strength of the CMB temperature fluctuation as a function of angular scale, varying the geometry of the Universe and the relative fraction of normal matter and dark matter between the models. The technical term for a plot of the strength of the temperature fluctuations as a function of angular scale is the "*CMB power spectrum*".

One of the theoreticians who contributed to this work was Dick Bond of the Canadian Institute for Theoretical Astrophysics. Bond was born in Toronto Canada in 1950, and did his first degree in physics at the University of Toronto, graduating in 1973. He then went to Caltech to do his Ph.D. in theoretical physics, his Ph.D. supervisor was Bill Fowler, who as we saw in Chap. 3 had worked with Fred Hoyle in the 1950s on nucleosynthesis within stars. By the mid 1980s, Bond was a professor at the Canadian Institute for Theoretical Astrophysics in Toronto. In a series of papers in the mid 1980s, Bond, Jim Peebles, George Efstathiou of Cambridge University in England and others developed the theories of what could be learnt about the properties of the Universe from studying the CMB's power spectrum, e.g. [3–7].

They argued that, whilst at larger scales the number of fluctuations at different angular scales should be roughly constant (what we would call a "flat power spectrum"), this was not true at smaller angular scales. By the time one reached angular scales of about 1° and smaller (1° corresponds to about twice the angular size of the full Moon), there should be a number of separate peaks in the power spectrum. The position and relative heights of these peaks, they argued, could tell us about the geometry of the Universe, its expansion rate, its matter density, the ratio of ordinary matter to dark matter and other basic parameters which astronomers had been struggling to measure for decades to pin down.

As an example, the models predicted that if the geometry of the Universe were flat, as inflation theory argued it was, then as we moved from the smallest angular scale observed by COBE of 7° to even smaller angular scales, there should be a large peak in the CMB power spectrum at 1°. If the Universe's geometry were curved rather than flat, the theory predicted that the most common angular size of the fluctuations would be different. An open Universe with negative curvature would have the most common fluctuations occurring at an angular scale larger than 1°, and if the Universe were closed and had positive curvature then the most common fluctuations would occur at angular scales smaller than 1°. An example of this is shown in Fig. 5.1.

5.2 Looking for the First Acoustic Peak

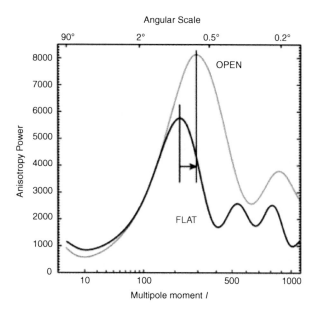

Fig. 5.1 A typical theoretical curve of the CMB power spectrum. The x-axis is the angular scale, and the y-axis is the power (or temperature difference from the average) at that angular scale. This plot shows the relative position of the first acoustic peak for a flat and open geometry (image credit: NASA)

This peak became known as the "**first acoustic peak**". Once the excitement over COBE finding the larger scale anisotropies had died down, experimentalists quickly turned their attention to devising experiments to see whether this first acoustic peak existed or not, and whether its position would correspond to a flat, closed or open Universe.

5.2 Looking for the First Acoustic Peak

The first observations of small scale structure in the CMB were made in 1995 by the Cosmic Anisotropy Telescope, a three element radio interferometer operating at the Mullard Radio Astronomy Observatory, part of the University of Cambridge [8]. This small array observed the CMB from 13 to 17 GHz with angular scales from about 0.25° to just over 0.5°. At these small scales they found temperature fluctuations of about two parts in 10,000, consistent with the anisotropies found by COBE at larger angular scales. The observing community debated whether investigating the small scale structure of the CMB could be done from ground-based and balloon-borne experiments, or whether the observations would require developing a new satellite, something which would take the best part of a decade.

Dave Wilkinson, the Princeton veteran of CMB research, had been part of the DMR team on COBE which Smoot had led. He had not particularly enjoyed the experience of working on his first satellite project. He had been used to working with a small group of colleagues, mainly based on Princeton, and the previous experiments that he had worked on had been designed, fabricated and tested at Princeton before deployment. He found working in the large COBE DMR team to be an entirely difference experience. All of the fabrication, testing and assembly of the instrument components had been done at Goddard, and with dozens of scientists and engineers working on that one instrument, Wilkinson had found it to be a far less satisfying experience than his previous experiments.

Naturally then, he was reluctant to get involved in a new satellite project. But another Princeton physicist by the name of Lyman Page was able to persuade him that a new satellite was needed to properly investigate the details of the small scale structure of the CMB to the necessary level of precision. Page had done his undergraduate degree at the small college of Bowdoin in southern Maine in the early 1980s, majoring in physics. After completing his degree he made the rather unusual step of taking a job in Antarctica, based at the U.S. base of McMurdo Station, on the coast of this most southerly continent. Ever since the 1970s, radio astronomers had been heading to Antarctica to make observations which were just not possible anywhere else on Earth.

5.3 The Most Southerly Continent

Each winter, when I worked at Yerkes Observatory, the place would empty out for many months. Most of my colleagues who worked at the Observatory were "heading south", and when I say "south" I mean as far south as it is possible to go. They were heading to do vital maintenance work during the southern hemisphere summer on telescopes situated at the south pole. These telescopes were operated by the Center for Astrophysical Research in Antarctica (CARA), a large multi-university organisation headed up by my then-boss, Professor Doyal A. Harper of the University of Chicago.

For many centuries Antarctica had been a mythical continent, the "*terra incognita*" or "*terra australis*" (southern land) that The Royal Navy had instructed Captain Cook to search for when he had finished his observations of the Transit of Venus in 1769. Cook failed to find the land on this voyage, and he made two further voyages in the southern oceans, but failed to find this mythical continent on all three attempts. We now know that during his second voyage, his ships HMS Resolution and Adventure, when they crossed the Antarctic circle in January 1773, December 1773 and January 1774, came within about 120 km (75 miles) of the Antarctic coast, but they failed to see any land on each occasion. Each time they were prevented from heading further south by the increasing sea-ice as they got further and further within the Antarctic circle.

The first voyages to catch sight of the land which would become known as Antarctica were in 1820, a Russian expedition led by Fabian Gottileb von Bellingshausen and Mikhail Lazarev on the ships the *Vostok* and *Mirny*. They spotted the ice shelf of the continent, but did not land their ships on its coast. A few days later a Royal Navy ship captained by Edward Bransfield also spotted the Antarctic coast, and later in the same year Nathaniel Palmer, a sealer from Connecticut in the USA made the third sighting of the continent. There is some dispute as to when the first landing occurred. John Davis, another American sealer, claimed to have made a landing in February 1821 at a place known as Hughes Bay, but this claim is disputed by some. The next landing, which many people believe could have been the first landing, was made in 1895 at Cape Adair by three men from a Norwegian whaling expedition, Henryk Bull, Carstens Borchgrevink and Alexander von Tunzelmann. Whether John Davis actually landed or not in 1821, what is clear is that virtually no exploration of the Antarctic continent happened in the nineteenth century. All that was about to change.

On the 6th of August 1901 a Royal Navy ship called *Discovery* left the Isle of White, an island a few kilometres from Portsmouth off the south coast of England, bound for Cape Adair. Two of the men on the ship would become amongst the most famous Antarctic explorers of the twentieth century; the ship's Captain Robert Falcon Scott and the ship's Third Officer Ernest Shackleton. The Discovery arrived in New Zealand, via Cape Town, in the November, and headed on to Cape Adair, arriving in January 1902. They then sailed southwards and eastwards, eventually entering McMurdo Sound in early February, they anchored at a place at its southern limit which they later named Winter Quarter Bay. The expedition team built several huts on a rocky peninsula they named Hut Point, which was at a latitude of $77°47'$S.

The Discovery expedition was jointly organised by the Royal Society and the Royal Geographic Society. Amongst the ship's crew was Edward Wilson, who was travelling at the Expedition's junior doctor and zoologist. As Wilson later stated, one of the Expedition's stated aims, in addition to learning about the plant life, wildlife and geology of this new land, was

> to get as far south in a straight line on the Barrier ice as we can, reach the Pole if possible, or find some new land.

As a very minimum, they wanted to get closer to the South Pole than Norwegian Borchgrevink had done in February 1900, when he had reached a latitude of $78°50'$S.

The three who headed off to travel as far south as possible were Scott, Shackleton and Wilson. They left Hut Point on the 1st of November 1902, and passed Borchgrevink's latitude record on the 11th. They continued on, passing the 80° mark on the 2nd of December. On the 30th of December the expedition reached 82°17'S, but at this point had to return as they were running low on rations, and the health of all three men was suffering. On the return journey, Shackleton collapsed and was unable to pull the sleds. Once they had got back to Hut Point, Scott had Shackleton sent back to Britain on the expedition's relief vessel, an act which Shackleton hugely resented and the two fell out as a consequence, never to work together again.

A few years later Shackleton organised his own expedition to reach the South Pole. The *Nimrod Expedition* set sail in August of 1907 and arrived in McMurdo Sound in late January 1908. Shackleton had intended to use the huts that had been built for the Discovery Expedition, and in fact had written to Scott before leaving England to alert him of this. Scott had initially been reluctant, telling Shackleton that he too had plans to return to Antarctica, but after Wilson mediated between the now rivals, Scott acquiesced and gave Shackleton his permission. However, when the Nimrod arrived in McMurdo Sound in the January of 1908, Shackleton found they were unable to land safely near Hut Point, and so eventually they landed at Cape Royds, about 40 km (25 miles) from Hut Point. It is here that they put up their prefabricated hut and made plans for an assault on the Pole.

On the 29th of October 1908, Shackleton and three others started their trek for the South Pole. On the 26th of November they passed the 82°17'S record that Shackleton had set with Scott and Wilson in December of 1902, and moved into unknown territory. In early December, they found themselves at the base of a vast glacier which they named the Beardmore Glacier, after the expedition's main sponsor. It took them the best part of a month to ascend this glacier, they finally emerged on to what we now call the Antarctic plateau on the 26th of December, but with their food rations running low. They had left food in depots along the way, but had barely a month's worth of rations with them. On the 4th of January 1909 Shackleton admitted defeat, they would not be able to get to the Pole with the rations they had remaining. On the 9th of January they made a last march without the sledges or other equipment, and managed to reach a latitude of 88°23'S. This was only 97 miles (about 160 km) from the Pole, and Shackleton had to accept the smaller achievement of getting within 100 miles of their target.

In replying to Shackleton's request to use the huts at Hut Point, Scott had mentioned his own intentions of returning to Antarctica, but had not been willing to divulge any details. Scott was putting together his own expedition, which would become known as the "*Terra Nova Expedition*". The Terra Nova set sail from Cardiff on the 15th of June 1910, and Scott had dined at Cardiff's Royal Hotel on the evening before the Expedition's departure. There are several memorials to this to be found in Cardiff today, including a plaque on the outside wall of the Royal Hotel (see Fig. 5.2), the light house on Roath Park lake (see Fig. 5.3), one of the city's parks; and on one of the staircases in Cardiff City Hall there is a memorial to Scott and his companions (see Fig. 5.4). Scott himself did not sail with the Terra

5.3 The Most Southerly Continent

Fig. 5.2 The Royal Hotel, where Scott dined on the 13th of June 1910, 2 days before the *Terra Nova* set sail from Cardiff (image credit—the author)

Nova from Cardiff, as he was detained in Britain making final arrangements for the Expedition. He sailed on a faster ship and joined with the Terra Nova in South Africa. In Melbourne Australia Scott re-engaged in some further fundraising for the Expedition, whilst the Terra Nova sailed on to New Zealand.

Fig. 5.3 The light house at the southern end of Roath Park lake in Cardiff, erected to commemorate Scott's *Terra Nova* expedition to the South Pole sailing from Cardiff in 1910. On the *right* are the details of the inscription on the plaque (image credit—the author)

Fig. 5.4 A plaque in Cardiff City Hall which commemorates Scott's Terra Nova sailing from Cardiff in its journey south (image credit—the author)

5.4 "Great God! This Is an Awful Place"

It was also in Melbourne that Scott found a telegram waiting for him from renowned Norwegian Arctic explorer, Roald Amundsen. Before leaving Britain, Scott had heard that Amundsen was planning an expedition, but assumed it would be another attempt by the Norwegian to explore the North Polar regions. In 1903 Amundsen had led the first expedition to successfully navigate the Northwest Passage which linked the Atlantic and Pacific oceans in the northern parts of Canada. With the prize of being the first to reach the North Pole still unclaimed, Amundsen tried to raise funds to mount an expedition to be the first to get there. However, in 1909 he learned that two American expeditions led by Frederick Cook and Robert Peary had reached the North Pole by two different routes. He decided to divert his attentions instead to being the first to reach the South Pole.

The telegram from Amundsen awaiting Scott in Melbourne simply read

BEG TO INFORM YOU FRAM PROCEEDING ANTARCTIC. AMUNDSEN.

Fram was the name of Amundsen's ship, and Scott now knew he was in a race against the Norwegian to be the first to get to the South Pole. Scott rejoined the Terra Nova in New Zealand, and the ship arrived in McMurdo Sound in January of 1911. They decided to make their base at a cape about 25 km (15 miles) to the north of Hut Point, his base in the 1903 Discovery Expedition. He named the cape *Cape Evans* after his second in command, Lieutenant Edward Evans, and the team set about erecting a hut to house themselves and their equipment. This prefabricated hut was ready for the men to move into by the 18th of January.

Meanwhile, Amundsen and his team had sailed south on the Fram, also reaching Antarctica in early January 1911. Amundsen decided to land in the Bay of Whales, which was over on the other side of the Ross Sea and the Ross Ice Shelf from where Scott had landed (see Fig. 5.5). Part of the Terra Nova Expedition's plans was to explore the land to the east of where they had landed, so towards the end of January a party led by Victor Campbell set sail in the Terra Nova. After failing to land on the shores of the region which had been named King Edward VII land, they were returning westwards back to McMurdo Sound when they came across Amundsen and his ship in the Bay of Whales. Scott received this news on the 22nd of February, confirmation that his rival in the race for the Pole had also arrived.

Scott began his trek to the South Pole from Cape Evans on the 1st of November 1911. The party began their ascent of Beardmore Glacier in early December. By the 20th of December they had reached the beginnings of the Antarctic Plateau. On the 3rd of January 1912, they had reached a latitude of $87°32'S$, and it was on this day that Scott made the decision as to who in the party would continue on to the Pole. He chose four men to accompany him; the four being Henry Bowers, Edgar Evans, Lawrence Oates and finally Edward Wilson, the last of whom had accompanied Scott and Shackleton on their attempt to reach the Pole in 1902. The others in the party returned to Cape Evans, laying deposits of food as they made their way back for the Polar party to find on their return from the Pole. The group of five

Fig. 5.5 The Antarctic continent, showing the position of the South Pole and where Amundsen and Scott made their bases (*red square* and *black circle* respectively)

continued on to the Pole, and on the 9th of January they passed 88°23'S, the record set by Shackleton's party in 1909. They continued on, by now feeling certain that they would reach the South Pole. On the 16th of January 1912, within just 25 km (15 miles) of their goal, the party spotted the most heart breaking sight they could have imagined—through the cold clear air they could unmistakably see Amundsen's black flag in the distance. Knowing that they had lost the race, the following day, the 17th of January, they trekked the final 25 km to the South Pole, where they marked their arrival by planting the British Union flag and taking a photograph where each of their faces shows the agony of their defeat—see Fig. 5.6. Scott wrote in his diary

Great God! This is an awful place

On the following day, the 18th of January, Scott's party found Amundsen's tent with some supplies, a letter addressed to King Haakon VII of Norway, and a note from Amundsen asking Scott to deliver the letter. In the note, Amundsen told Scott that his party had arrived at the Pole on the 16th of November 1911, some 34 days before Scott's party's arrival. In a mood of utter dejection and despair, Scott and his four companions set off on the sad journey back to Cape Evans.

During the long trek back along the Antarctic Plateau, Scott began to worry more and more about the physical condition of his party. The most worrying to Scott were the conditions of Evans and Oates. Evans was suffering from severe frostbite, and had fallen several times taking severe blows to the head. Oates was also suffering badly from frostbite, particularly in his feet. This impacted markedly on his ability to pull the sledges that were carrying the party's supplies, and progress became slower and slower as Oates struggled to make progress.

5.4 "Great God! This Is an Awful Place"

Fig. 5.6 Scott and his colleagues at the South Pole. They arrived on the 17th of January 1912, only to find that Amudsen and his party and got there 34 days beforehand on the 16th of November 1911. The look of utter failure and despair is evident on the faces of the men

On the 7th of February they began their descent of the Beardmore Glacier, but by this time Evans's health was deteriorating even further. In addition to his severe frostbite and head injuries, he had also badly cut his hand during one of his falls and it was failing to heal. During the descent of the glacier Evans got weaker and weaker, and near the bottom of the glacier, on the 17th of February, he collapsed and died.

Once the party had reached the base of the glacier, they only had the home stretch to Cape Evans to do, along the barrier ice. However, the weather conditions got worse, with cold temperatures and high winds. As the men got weaker and weaker from lack of food and the terrible conditions, their progress became slower and slower. The cold weather made pulling their sleds even harder, the temperatures were often too low for the runners to form a layer of melted ice between the sled and the barrier ice, which made pulling the sleds "like pulling over desert sand" Scott would write in his diary on the 19th of February.

Oates' foot continued to get worse, and he became increasingly unable to walk. As February moved into March, the weather became even colder, and the daily progress of the party slowed so much that they were barely covering 8 km (5 miles) each day. This made their food situation even worse, as it took longer between reaching each new supply of food and fuel which had been laid down for their return journey. To make matters even worse, on the 2nd of March at one of the depots, Scott found that the ration of fuel was depleted, apparently a result of evaporation during storage.

They encountered the same shortage of fuel at the next depot on the 9th of March, but pushed on. Oates became more and more delirious, and it was clear to him and the others that his condition was slowing the party's progress. On the 17th of March, whilst the party were sheltering from a snow storm, Oates said to the others

> I am just going outside and I may be some time.

and he walked off to die in the blizzard, hoping that his death may allow the others to improve their rate of trekking and to reach the next depot before their current supplies ran out. The next depot, known as *One Ton Depot*, was the largest supply of food and fuel which had been laid down, and reaching it should allow them to restock their supplies sufficiently to be able to make it back to Cape Evans.

The remaining three, Scott, Wilson and Bowers pushed on, but just 3 days later on the 20th of March they encountered a fierce blizzard and were forced to pitch their tent and take shelter until the weather improved. Each day they tried to advance on to One Ton Depot, which they knew was only about 16 km (10 miles) away, a distance they could make in 2 days if lucky. But each day they were unable to advance due to the severity of the weather. Just over a week later, all three perished in their tent. Scott's last diary entry is on the 29th of March, where he wrote

> We shall stick it out to the end, but we are getting weaker, of course, and the end cannot be far. It seems a pity but I do not think I can write more. R. Scott. Last entry. For God's sake look after our people.

5.5 The Center for Astrophysical Research in Antarctica

Some of the reasons that Scott and his party perished on their return from the South Pole was due to the severity of Antarctica's climate. Antarctica is the most inhospitable of the World's continents. It is the coldest, driest continent; it holds the World record for the coldest temperature ever recorded ($-89°$ C) and has an annual precipitation of less than 200 mm, which technically makes it a desert. The Antarctic Plateau, on which the South Pole lies, experiences much less precipitation than this. The plateau stretches over a diameter of about 1,000 km (620 miles), and has an average elevation of about 3,000 m (9,800 ft). However, because of the very cold temperature, the atmosphere is actually thinner at the South Pole, and the 3,000 m elevation feels more like 4,500 m or so.

The elevation of the Antarctic plateau is due to the very slow accumulation, over millions of years, of snow which has fallen on the frigid interior of this continent. Some 2 km below the surface of the plateau is solid land, the South Pole literally sits on top of a 2 km thick ice sheet. There is no naturally occurring life on the Antarctic Plateau, even at the microbial level, but despite these inhospitable conditions there has been a continuous human presence at the South Pole since 1956.

In November 1956 the US Government built the Amundsen-Scott South Pole Station as part of its commitment to the International Geophysical Year, which ran from the 1st of July 1957 until the 31st of December 1958. The base established by the US Government in 1956 has been continuously occupied by scientists and support staff ever since. This was the base that Smoot and his colleagues from

Berkeley were able to use during November and December 1981 when they headed to the South Pole to deploy a radio dish to measure the microwave emission of the Milky Way galaxy as part of their verification of the anisotropies that COBE was measuring (see Sect. 4.8).

It became obvious from their observations and those made by other teams that Antarctica offered a site almost unequalled in the World for making observations in the millimetre and microwave parts of the spectrum. Whereas at more temperate latitudes such observations were hampered by emission from water vapour in the Earth's atmosphere, the extreme cold and dryness of the atmosphere above Antarctica made it possible to make such observations with significantly lower emissions from the atmosphere.

By the late 1980s it was clear that US astrophysical research in Antarctica would be best served if there were an organisation to coordinate the efforts of a number of universities and research organisations. A proposal was submitted to the National Science Foundation, and in 1991 the Center for Astrophysical Research in Antarctica (CARA) was established, with my ex-boss Doyal A. Harper as its first Director. CARA would deploy telescopes at an area a few kilometres away from the Amundsen-Scott Station, at a place which has become known as the "*Dark Sector*", and use the facilities at the Station to provide living accommodation and support for the engineers and astronomers working on the telescopes.

Being at the South Pole also provides other unique challenges and benefits. Most of us living in latitudes outside of the Tropics are very familiar with the changing length of the day as we move from Winter Solstice (around December the 21st in the Northern Hemisphere) to the Spring Equinox (around the 21st of March) and on to the Summer Solstice (around the 21st of July). The difference in the length of the day between the Winter and Summer Solstices in a city like London, which is at a latitude of $51°32'N$, is a factor of about two. The time between sunrise and sunset on the day of the Winter Solstice is about 8 h, on the day of the Summer Solstice this has stretched to about 16 h.

This difference becomes more and more extreme as one heads further and further from the Earth's equator. The Arctic and Antarctic circles lie at latitudes of $66°30'$ North and South of the Equator respectively. If one is at the Arctic circle, then on the day of the Summer Solstice, the Sun will not actually set but will stay above the horizon all "day". Further north than this, and there will be more than 1 day when this is true. Finally, at the North Pole itself, there is only one sunrise and one sunset a year! On the Spring Equinox the Sun will rise (because of atmospheric bending of sunlight, this actually happens a few days before the day of the equinox), and it will stay above the horizon until a few days after the Autumn Equinox, when it will finally set. A day at the North Pole actually lasts one whole year, with 6 months of continuous daylight followed by 6 months of continuous darkness.

What is true for the North Pole is also true for the South Pole, except that it works in the opposite direction. In March, when the Sun is rising at the North Pole, it is setting at the South Pole marking the beginning of the long South Pole night. The Sun will stay below the horizon of the South Pole until just before the equinox in September, when it will finally rise again as it sets at the North Pole. A

6-month long night has many advantages for astronomical observations, allowing continuous monitoring of objects in a way which is not possible from any other location on Earth. Objects outside of the Solar System, including the centre of our Milky Way galaxy, remain at the same elevation above the horizon continuously, making tracking them much easier than at other latitudes.

However, because of the 6-month darkness and the even colder temperatures during the Austral winter, getting in and out of the South Pole presents unique challenges. The Amundsen-Scott base is operated by the US National Science Foundation (NSF), and getting to the South Pole involves first of all flying on an NSF aeroplane from Christchurch New Zealand to McMurdo Station in McMurdo Sound, and then a separate flight onto the South Pole. Neither flight is comfortable. The long flight from Christchurch to McMurdo is undertaken in a C-141 military transport aeroplane, stripped bare of any niceties like comfortable seats or sound and heat insulation. Passengers board in the relative warmth of a New Zealand summer dressed in the thermal clothes they will need for when they disembark in McMurdo. The long flight from Christchurch to the edge of the Antarctic continent is one of the most dangerous flights a person could undertake, possibly second only to the flight from McMurdo to the South Pole itself.

Before boarding, the scientists, engineers and technicians heading to McMurdo are left in no doubt of the dangers awaiting them. Should the aeroplane go down in the Southern Ocean there is essentially no chance of being rescued as the survival time in the frigid waters is too short to allow a rescue aeroplane to arrive. Passengers are told that in such an eventuality they have two choices, the first is to keep their heads above the water and die an agonising death as the organs in their bodies shut down one by one in the freezing water. The second option would be to make the agony less drawn out by submerging their heads, whereupon death is much quicker as the brain will shut down very quickly in the frigid water. Faced with such danger, it is a wonder that any people are even willing to undertake the flight, but actually the NSF has never suffered a ditch into the ocean so no one has yet had to employ any of these morbid instructions.

Once the C-141 has arrived in McMurdo, and after a few days' acclimatisation, those people going on to the South Pole will board another military transport aeroplane, this time a C-130, to undertake the flight over the Antarctic Plateau to the Amundsen-Scott South Pole Station. This aeroplane is not equipped with wheels but rather skis, and lands on a runway which digger lorries from the South Pole Station keep clear and smooth on a daily basis. So cold are the temperatures at the South Pole that the aeroplane has to keep its engines running, if it were to switch them off the fuel lines would freeze and there would be no chance of getting them started again. When the passengers step out of the aeroplane they are greeted by a sight unlike any they will have seen before—a vast expanse of whiteness as far as the eye can see in every direction, except for a few nearby buildings in the direction of the South Pole station which represent the only human presence for over 1,000 km.

People travelling to the Amundsen-Scott South Pole Station fall into two categories, those who will be leaving before February, and those who will not. This is because in February all NSF flights to the South Pole cease, the temperatures become too cold for the aeroplane to operate. From February until October, a period of some 8 months, the base is essentially cut off from the outside World, with no flights in or out of the station. The people who remain there after the base closes to the outside World are known as *"winter overs"*, and they face 8 months of physical isolation from the rest of the World, although thankfully they can stay in internet contact for parts of each day thanks to a few errant communication satellites which pop up above the horizon for several hours each day.

By late March the Sun has set and the Station is plunged into a 6-month long darkness, the Sun won't reappear until just before the September equinox. The South Pole Station typically has about fifty people who winter over, most of them are scientists working on various research projects, but there are also support crew to help keep the station running, and a doctor in case of medical emergencies. Possibly the most important member of staff is the cook, and the food is both plentiful and very tasty. At the extreme cold of the South Pole winter the body burns up far more calories than normal in an attempt to keep warm. If you add to that the fact that most of the scientists are involved in projects which require skiing out to their experiments, taking measurements and collecting things, and a typical scientist wintering over can easily burn up about twice the amount of calories that you and I do, and sometimes considerably more. Large amounts of food is therefore essential to keep the scientists from falling foul of possible malnutrition, but the richness and tastiness of the excellent food has the added bonus of keeping morale up amongst the winter overs during their long months of isolation.

The only respite to the daily monotony occurs around late June, mid-winter at the South Pole. Around this time the NSF organises an air drop of items for each person at the Pole, with some fresh fruit and vegetables for the whole station, and letters and even gifts from loved ones back home. From the astronomers who have wintered over at the Amundsen-Scott South Pole Station I have heard that this airdrop provides some welcome excitement during the long months of darkness, cold and hard work. One of the other things which lifts the spirits during these months of blackness are the frequent auroral displays.

The southern (and northern) lights, *aurora australis* and *aurora borealis* respectively, are caused by electrons which have been blown from the Sun's surface hitting the Earth's magnetic field and raining down through the Earth's atmosphere towards the magnetic poles (Fig. 5.7). The eruptions which lead to the electrons leaving the Sun's surface are known as solar storms, and are more frequent when the Sun has more sunspots. There is always a steady stream of charged particles coming from the Sun, this is known as the Solar wind, but during a particularly energetic storm large amounts of charged material can erupt from the surface in an event known as a *"coronal mass ejection"*. If this ejection happens on part of the Sun facing the Earth, then the charged particles will come hurtling towards Earth, taking about 3–4 days to traverse the 150 million km.

When they reach the Earth the electrons get trapped and funnelled along the Earth's magnetic field lines, and they stream towards the north and south magnetic poles, travelling down through the atmosphere. These charged electrons will excite the electrons in the nitrogen and oxygen molecules in the Earth's atmosphere, and

Fig. 5.7 A display of the aurorae australis (southern lights) with the South Pole 10 m Telescope in the foreground (image credit—NSF)

cause those molecules to glow. The colours we see are due to the different emission lines in the nitrogen and oxygen, with the ever changing colours and appearance being produced as different energy levels are excited and different parts of the atmosphere are excited.

The first CMBR telescope which CARA deployed at the South Pole was called the White Dish, a 1.4 m diameter radio dish which operated from 1991 until January of 1993. It made observations at 90 GHz with a beam size (the term radio astronomers use for the resolution of a radio telescope) of 12′ [9]. Overlapping in operation with White Dish was Python, a 0.75 m diameter radio dish which operated for five Austral summers from 1992 until 1997. In its first three summers it observed at 90 GHz, the same frequency as White Dish, and then in its final two summers it observed at 40 GHz [10]. Python was superseded by VIPER, a 2.1 m radio dish which went into operation in 1998. The final telescope in CARA's 10-year lifetime was DASI, the Degree Angular Scale Interferometer, which went into operation in 1999 and operated until CARA's end in 2001.

During its 10 year lifetime, CARA's operations were coordinated from Yerkes Observatory, with my colleague Bob Pernic being in charge of deciding who would be at the South Pole Station at what time and for how many weeks. As well as coordinating the efforts of many of my colleagues who were based at Yerkes and at the University of Chicago's campus in the city, Bob also coordinated the work of the other astronomers and technicians across the dozen or so universities and institutions which were part of CARA project. During the months from October to February, during the months of continuous daylight at the South Pole, as many astronomers and astronomy engineers as the Amundsen-Scott Station could accommodate would head down to the Pole for a 2 or 3 week tour of duty to work on repairing or upgrading the telescopes ready for the following winter's observations.

In late January or early February the scientists who had been chosen to winter over with the experiments would head down, and by mid February the Station would empty out as the last few flights took the people who were not wintering over back to McMurdo and onwards to their home universities, leaving the 50 or so hardy men and women isolated at the South Pole until the Station reopened to the outside word in October. Whilst the CARA telescopes Python, and later VIPER, were starting to make important observations of the small scale structure of the CMB from the South Pole, other astronomers had returned to using balloons to make their observations.

5.6 The Universe is Flat!

In 1996 a balloon-borne CMB experiment called QMAP flew twice from the U.S. Balloon Launch Facility in Palestine Texas. It made observations of the anisotropies in a small part of the sky, giving results that were consistent with the DMR map produced by COBE a few years earlier [11]. Over the following 2 years, in 1997 and 1998, the gondola from the QMAP experiment was used by the Mobile Anisotropy Telescope (MAT) to make more detailed observations of fluctuations in the CMB at three different frequencies, 30, 40 and 144 GHz [12, 13]. MAT was located in Cero Toco in Northern Chile and, although it was ground-based rather than balloon-borne, the 5,600 m (over 18,000 ft) elevation of Cero Toco allowed a level of sensitivity sufficiently high to measure fluctuations in the CMB at these several frequencies and at angular scales ranging from about 3.5° down to 0.9°. MAT/TOCO became the first experiment to make measurements of both sides of the 1st acoustic peak, and argued that the peak itself was at an angular scale of $0.83 \pm 0.1°$ [13].

The following year, in 1999, another balloon-borne experiment called BOOMERanG made even more precise observations of the position of the peak. After being tested in 1997, the 1.3 m telescope was launched on the 29th of December 1998 from the McMurdo Station in Antarctica, and was carried high in the cold Antarctic air by the Polar vortex around the South Pole. The balloon achieved an altitude of just over 36 km (about 120,000 ft), and stayed aloft for 10.5 days, slowly moving in an anticlockwise direction around the Pole, and making observations of the CMB at frequencies of 90, 150, 240 and 400 GHz and at angular scales from about 4.5° down to just over 0.15° [14]. Figure 5.8 shows the BOOMERanG balloon just before launch, and the high-quality anisotropy map of a small patch of the sky which it produced (the COBE DMR picture is also shown for comparison to highlight not only the small patch of the sky which BOOMERanG was able to observe, but also the much higher angular resolution of its observations).

The analysis of the size distribution of the fluctuations observed by BOOMERanG gave confirmation that the geometry of the Universe was flat. The importance of the result was not lost on the popular press, in April 2000 the New York Times ran the headline [15]

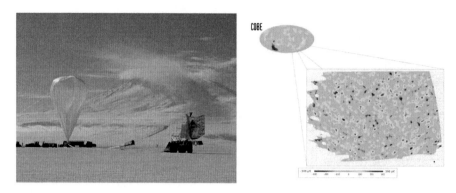

Fig. 5.8 BOOMERanG being readied for launch in Antarctica with Mount Erebus is in the background. On the *right* is the map of the CMB which BOOMERanG produced, compared to the COBE image, showing the much higher resolution of the BOOMERanG experiment (image credit—NSF/BOOMERanG)

Clearest Picture of Infant Universe Sees It All and Questions It, Too

The story explained that BOOMERanG had confirmed one of the major predictions of inflation theory, that the Universe is flat, but it also pointed out that the experiment had failed to see the smaller scale peaks which theorists predicted lay beyond the first acoustic peak. Either the theory of their prediction was wrong, or more sensitive experiments would be needed to see them.

Amongst several telescopes in the hunt for the smaller peaks predicted at angular scales less than the first acoustic peak was the Cosmic Background Interferometer (CBI), located again high in the Chilean Andes at a site called Llano de Chajnantor Observatory, at an elevation of just over 5,000 m (nearly 17,000 ft). CBI was a multinational project led by Caltech, and went into operation in 1999. It consisted of thirteen 0.9 m dishes, and operated at a range of ten frequencies from 26 to 36 GHz in 1 GHz steps. The interferometer allowed for high angular scale observations down to just under 0.1°. The CBI was the first experiment to find and measure the 2nd and 3rd acoustic peaks of the CMB power spectrum, and was also the first experiment to detect fluctuations down to angular scales corresponding to the seeds of the clusters of galaxies we see in the Universe today [16, 17].

5.7 Polarisation in the Cosmic Microwave Background

Meanwhile, back at the South Pole, CARA's final telescope in its 10-year lifetime, the DASI telescope, went into operation in 1999. DASI (Degree Angular Scale Interferometer) was also attempting to study the angular scale and strength of the 2nd and 3rd acoustic peaks in the CMB power spectrum. But, unlike the CBI experiment, DASI was also fitted with a polarimeter which would allow it too look

5.7 Polarisation in the Cosmic Microwave Background

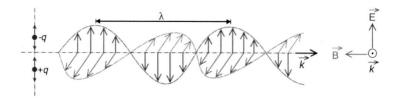

Fig. 5.9 An Electromagnetic (EM) wave. The *blue wave* is the electric field, the *red wave* is the magnetic field. An EM wave self propagates through space, and travels in a vacuum at the speed of light c

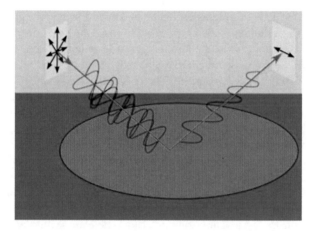

Fig. 5.10 When light is reflected off of a surface light a lake, it becomes polarised. The reason for this is that it is easier for the electrons in the water molecules in the plane of the surface of the lake to be jiggled by the incoming photons than it is for those perpendicular to the surface

for any polarisation in the CMB's light. Electromagnetic (EM) radiation travels through space at the speed of light, and the wave is composed of a varying electrical field at right angles to a varying magnetic field (see Fig. 5.9). The varying electrical field produces a varying magnetic field, and the varying magnetic field produces a varying electric field. In this way, the EM wave sustains itself as it travels through space, it is the ultimate in self-sufficiency!

Normal light (and radiation at other wavelengths outside of the visible) tends to be unpolarised, which means that the direction of the electric field (and the magnetic field which is at right angles to it) can be orientated in any direction. However, certain processes cause light to become polarised, and one of these is when light is reflected. Figure 5.10 illustrates this for light being reflected off the surface of a lake. The incoming rays have random polarisations, but the reflected rays have been polarised by the interaction with the surface of the lake. The electric-field directions which are not parallel to the surface of the lake are not reflected, or reflected at a greatly reduced level. The only orientation which is reflected efficiently is the one which is parallel to the water's surface, and so the reflected light is polarised in this direction.

A polarising filter is a filter which only allows light with a particular polarisation to pass through, such filters are found in polarising sunglasses. Even without any reflections off lakes and other reflective surfaces, glasses with polarising filters will cut down on the intensity and glare from light in the summer because they only allow the light with a vertical polarisation to pass through. But, they are even more effective around horizontal reflective surfaces like lakes and the sea, as the light reflected from such surfaces will be polarised parallel to the surface, with no reflected light having a vertical polarisation, the only light the lenses will allow through. In this way, such sunglasses significantly cut down on the glare.

In astronomy, radiation is also typically polarised by scattering. For example, when starlight scatters off interstellar dust grains in reflection nebula, the starlight which reaches us is not only preferentially blue due to more blue light than red light being scattered, but it is also polarised. The reason the sky is blue is because of the Rayleigh scattering of light off of the electrons in the atmosphere. Again, it is the blue light which is scattered more than red (about 16 times more!), and this makes the sky look blue. The reddening of the Sun as it sets is due to the same effect, but this time the scattering of the blue light as it passes through more and more of the atmosphere as the Sun gets closer to the horizon removes it from our line of sight, leaving the red light to come through, which makes it look redder.

In the CMB, the photons which reach us after they decouple from the matter are scattered off of the free electrons just before they recombine with the hydrogen nuclei, something known as Thomson scattering. Thomson scattering is named after J.J. Thomson, who was the first Director of Cambridge University's Cavendish Laboratory and who discovered the electron in 1987. Theoreticians had predicted two types of polarisations in the CMB, a so-called "*E-mode*" due to the density fluctuations in the CMB at the moment of decoupling, and a lower level "*B-mode*" which would be due to gravitational waves in the very early Universe. We shall return to the B-mode polarisation in Chap. 7.

5.7.1 How Does E-Mode Polarisation in the CMB Arise?

The E-mode polarisation arises from Thomson scattering of photons off of the electrons in the CMB just before the electrons combined with the nuclei, so at the moment of decoupling of matter and radiation. Let us suppose we have unpolarised light just coming from one direction, when this scatters via Thomson scattering off of an electron it is polarised. This is shown in Fig. 5.11.

Of course, when the CMB was formed we did not have photons from just on direction hitting electrons, they would have come from all directions. Let us suppose now that we have two photons scattering off of an electron at the same time, coming from different directions, as shown in Fig. 5.12. One electron comes in from the left, the other from the top, and after scattering off of the electron each is individually

5.7 Polarisation in the Cosmic Microwave Background

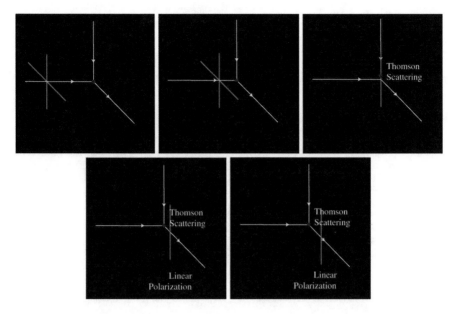

Fig. 5.11 If we have unpolarised light coming from just one direction then, after Thomson scattering off of an electron, it will be polarised. In this example it is scattered through 90° towards us (Images courtesy Wayne Hu, University of Chicago)

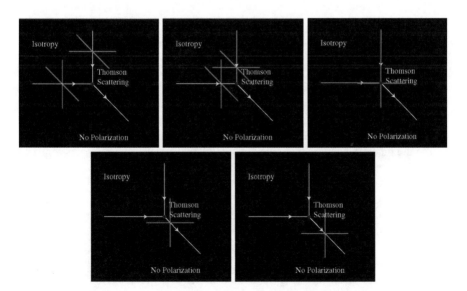

Fig. 5.12 If we have light *of equal intensity* coming in from the top and the left, then although each is polarised, the light scattered is not because it is the combination of the two polarised waves, but in combining the scattered wave is unpolarised (images courtesy Wayne Hu, University of Chicago)

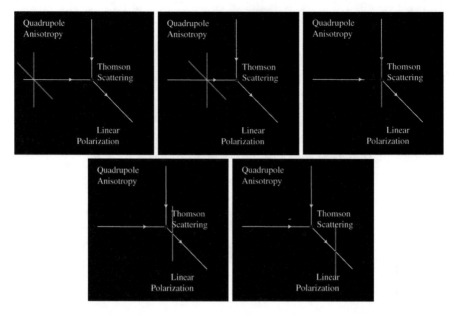

Fig. 5.13 If we have light of *different* intensity coming in from the top and the left, then the scattered light has a net polarisation as the *blue vertical field* is stronger than the *red horizontal* one, so we would measure a polarisation in the vertical direction (images courtesy Wayne Hu, University of Chicago)

polarised, but the scattered beam of light is a combination of the two, and so there is no overall polarisation.

Unpolarised scattered light, as shown in Fig. 5.12 would occur if electrons were being hit by photons from different directions which had the same intensity, but this was not the case. Because of the density variations in the gas, there were not areas and cold areas, these of course are the hot and cold spots in the CMB. Photons from the hotter areas would have a higher intensity than photons from the colder areas, and so the incoming electric field vectors would not have the same size. As Fig. 5.13 shows, if we now consider the same thing as illustrated in Fig. 5.12, but with electric field vectors of different sizes (the blue ones coming from the left are bigger than the red ones coming from the top), the scattered light will appear polarised as the blue vector and red vector are not the same size, so it will appear that the scattered light is polarised in the vertical (blue) direction.

DASI was looking for this E-mode polarisation. After confirming the positions and strengths of the first, second and third acoustic peaks in April of 2001 [18], it went is search of the never seen polarisation, which was predicted to be at a level of a few micro Kelvin. In December of 2002 they DASI team announced the successful detection of this E-mode in a paper in *Nature* [19] entitled *"Measurement of polarization with the Degree Angular Scale Interferometer"*. This was yet stunning vindication of a prediction of the big bang theory being observed. But, despite the numerous successes of DASI, BOOMERanG, CBI and other ground-based and balloon-borne experiments, cosmologists knew that the definitive observations would soon come from the Microwave Anisotropy Probe, which would be launched in June of 2001 but not produce its first all-sky map of the anisotropies until some 2 years later.

References

1. Peebles, P.J.E., Yu, J.T.: Primeval adiabatic perturbation in an expanding Universe. Astrophys. J. **162**, 815–836 (1970)
2. Sunyaev, R.A., Zel'dovich, Y.B.: Small-scale fluctuations of relic radiation. Astrophys. Space Sci. **7**, 3–19 (1970)
3. Bond, J.R., Silk, J., Kolb, E.W.: The generation of isothermal perturbations in the very early universe. Astrophys. J. **255**, 341–360 (1982)
4. Peebles, P.J.E.: Large-scale background temperature and mass fluctuations due to scale-invariant primeval perturbations. Astrophys. J. **263**, L1–L5 (1982)
5. Bond, J.R., Efstathiou, G.: Cosmic background radiation anisotropies in universes dominated by nonbaryonic dark matter. Astrophys. J. **285**, L45–L48 (1984)
6. Bond, J.R., Carr, B.J., Hogan, C.J.: Spectrum and anisotropy of the cosmic infrared background. Astrophys. J. **306**, 428–450 (1986)
7. Bardeen, J.M., Bond, J.R., Efstathiou, G.: Cosmic fluctuation spectra with large-scale power. Astrophys. J. **321**, 28–35 (1987)
8. Scott, P.F., Saunders, R., Pooley, G., O'Sullivan, C., Lasenby, A.N., Jones, M., Hobson, M.P., Duffett-Smith, P.J., Baker, J.: Measurements of structure in the cosmic background radiation with the Cambridge cosmic anisotropy telescope. Astrophys. J. **461**, L1–L4 (1996)
9. Tucker, G.S., Griffin, G.S., Nguyen, H.T., Peterson, J.B.: A search for small-scale anisotropy in the cosmic microwave background. Astrophys. J. **419**, L45–L48 (1993)
10. Platt, S.R., Kovac, J., Dragovan, M., Peterson, J.B., Ruhl, J.E.: Anisotropy in the microwave sky at 90 GHz: results from Python III. Astrophys. J. **475**, L1–L4 (1997)
11. Devlin, M., de Oliveira-Costa, A., Herbig, T., Miller, A., Netterfield, B., Page, L., Max Tegmark, M.: Mapping the CMB I: the first flight of the QMAP experiment. Astrophys. J. **509**, L69–L72 (1998)
12. Torbet, E., Devlin, M.J., Dorwart, W.B., Herbig, T., Miller, A.D., Nolta, M.R., Page, L., Puchalla, J., Tran, H.T.: A measurement of the angular power spectrum of the microwave background made from the high Chilean Andes. Astrophys. J. **521**, L79–L82 (1999)
13. Miller, A., Beach, J., Bradley, S., Caldwell, R., Chapman, H., Devlin, M.J., Dorwart, W.B., Herbig, T., Jones, D., Monnelly, G., Netterfield, C.B., Nolta, M., Page, L.A., Puchalla, J., Robertson, T., Torbet, E., Tran, H.T., Vinje, W.E.: The QMAP and MAT/TOCO experiments for measuring anisotropy in the cosmic microwave background. Astrophys. J. Suppl. Ser. **140**, 115–141 (2002)
14. de Bernardis, P., Ade, P.A.R., Bock, J.J., Bond, J.R., Borrill, J., Boscaleri, A., Coble, K., Crill, B.P., De Gasperis, G., Farese, P.C., Ferreira, P.G., Ganga, K., Giacometti, M., Hivon, E., Hristov, V.V., Iacoangeli, A., Jaffe, A.H., Lange, A.E., Martinis, L., Masi, S., Mason, P.V., Mauskopf, P.D., Melchiorri, A., Miglio, L., Montroy, T., Netterfield, C.B., Pascale, E., Piacentini, F., Pogosyan, D., Prunet, S., Rao, S., Romeo, G., Ruhl, J.E., Scaramuzzi, F., Sforna, D., Vittorio, N.: A flat Universe from high-resolution maps of the cosmic microwave background radiation. Nature **404**, 955–959 (2000)
15. Glanz, J.: Clearest picture of infant Universe sees it all and questions it. New York Times, 27 April 2000

16. Padin, S., Cartwright, J.K., Mason, B.S., Pearson, T.J., Readhead, A.C.S., Shepherd, M.C., Sievers, J., Udomprasert, P.S., Holzapfel, W.L., Myers, S.T., Carlstrom, J.E., Leitch, E.M., Joy, M., Bronfman, L., May, J.: First intrinsic anisotropy observations with the cosmic background imager. Astrophys. J. **549**, L1–L5 (2001)
17. Pearson, T.J., Mason, B.S., Readhead, A.C.S., Shepherd, M.C., Sievers, J.L., Udomprasert, P.S., Cartwright, J.K., Farmer, A.J., Padin, S., Myers, S.T., Bond, J.R., Contaldi, C.R., Pen, U.-L., Prunet, S., Pogosyan, D., Carlstrom, J.E., Kovac, J., Leitch, E.M., Pryke, C., Halverson, N.W., Holzapfel, W.L., Altamirano, P., Bronfman, L., Casassus, S., May, J., Joy, M.: The anisotropy of the microwave background to l = 3500: mosaic observations with the cosmic background imager. Astrophys. J. **591**, 556–574 (2003)
18. Halverson, N.W., Leitch, E.M., Pryke, C., Kovac, J., Carlstrom, J.E., Holzapfel, W.L., Dragovan, M., Cartwright, J.K., Mason, B.S., Padin, S., Pearson, T.J., Readhead, A.C.S., Shepherd, M.C.: DASI first results: a measurement of the cosmic microwave background angular power spectrum. Astrophys. J. **568**, 38–45 (2002)
19. Leitch, E.M., Kovac, J.M., Pryke, C., Carlstrom, J.E., Halverson, N.W., Holzapfel, W.L., Dragovan, M., Reddall, B., Sandberg, E.S.: Measurement of polarization with the degree angular scale interferometer. Nature **420**, 763–771 (2002)

Chapter 6
The Evidence Comes Together

Two large scale structure surveys catalogue the position and redshift of over 200,000 and 650,000 extra-galactic objects respectively, and in 1998 astronomers announce that the expansion of the Universe is accelerating. The cosmological constant and 'dark energy' are (re)introduced to explain this and other observations. By 2003 WMAP has produced the most spectacularly detailed measurements of the anisotropies in the cosmic microwave background from angular scales larger than COBE down to angular scales less than the smallest achieved from the ground. It pins down the values of cosmological constants like the age of the Universe, its density, the Hubble constant, and the percentages of baryonic matter, dark matter and dark energy to a level of accuracy unimagined a few decades ago.

6.1 Back into Space

Soon after COBE's successful detection of the cosmic microwave background (CMB) anisotropies, Lyman Page ran into Dave Wilkinson in the corridor of Princeton physics department. Page had done his Ph.D. at MIT, working with Steve Meyer's group on balloon-bore experiments to look for the fluctuations in the CMB. In the months before COBE announced its discovery of the long sought-after fluctuations in April 1992, Meyer and his team had found tantalising evidence for the anisotropies in their own data, but had held back from declaring a definitive discovery. The team just did not feel they could eliminate the possibility that the fluctuations they were seeing were not due to contamination from the atmosphere, the instrument, or from the Milky Way. When COBE scooped the headlines, Meyer's MIT team naturally felt envious that they had been beaten to the result.

After completing his Ph.D., Page moved to Princeton where he continued to be involved in balloon-borne experiments to map the small scale fluctuations in the CMB. He also got involved in the ground-based observations, like the ones being done high in the Atacama desert in northern Chile that I discussed in Chap. 5. However, these experiences made Page as aware as anyone of the limitations of trying to make such observations from somewhere other than space. The problem was that ground-based and balloon-bore observations could only map a small part of the sky, about 2° by 2°, an area about sixteen times the area of the full Moon. Any attempts to map a larger area of the sky were doomed to failure, because over larger areas there was just too much variation in the emission from the Earth's atmosphere, and these atmospheric variations were much larger than the tiny fluctuations in the CMB that they were searching for.

As we saw in Sect. 5.1, theory was predicting that the first acoustic peak in the CMB's power spectrum was at an angular scale of about 1°, with the subsequent peaks at smaller angular scales. Looking for the fluctuations at the smaller angular scales, although not easy, was possible from balloons and the ground, because in a 2° by 2° patch of sky the experiments could obtain enough independent measurements to do meaningful statistics. This is what CBI, BOOMERanG, DASI and MAT had done to varying degrees of success, but they would not be able to map larger areas.

To illustrate this, let us suppose an experiment is looking for fluctuations on an angular scale of 0.25°. In a 2° by 2° patch of sky this means it can make 64 independent measurements, as 0.25 will go into 2 eight times, and so into the area 64 times. Sixty four independent measurements is enough to say something meaningful about the statistics. But, in looking for fluctuations on the 1° angular scale, there would only be four independent measurements, which is far too few to be able to say anything meaningful.

These thoughts had been rattling around in Page's head for some time when he ran into Wilkinson and they started chatting about future CMB experiments. Page had never been involved in any satellite projects, and with his many years of doing balloon and ground-based work, did he really want to go through the learning process that a satellite mission would require? Wilkinson, on the other hand, had of course been one of the pioneers of CMB research. He and his Princeton colleagues were only a few months away from getting their experiment ready to look for the CMB when Penzias and Wilson made their discovery in 1964. He and Roll had obtained the second data point of the CMB's spectrum in December of 1965, in 1967 he and Partridge had shown the CMB to homogenous down to 0.1 %, and in the late 1970s he had become involved in the COBE satellite, working on the Differential Microwave Radiometer (DMR) experiment that eventually announced the discovery of the anisotropies in 1992.

Wilkinson's experience of working on COBE had not been a wholly pleasant one. He had a number of personal difficulties with the experiment's Principal Investigator (PI), George Smoot; but there were other reasons that he had not enjoyed the project. As an experimental scientist, he found it frustrating that the building and testing of the instrument's components was done wholly by engineers at Goddard. Wilkinson was the type of scientist who liked to get his hands dirty in the lab, who revelled in the joy of putting a complex instrument together and testing it, but all that joy was taken away from him on COBE.

Wilkinson also found it frustrating to work in such a large group spread across multiple institutes, he had been used to doing his work with colleagues in Princeton, all of whom worked on the same campus and often in the same building. On COBE, Wilkinson found himself part of a group which, at times, swelled to over one hundred people. Some people, he felt, used the size of the group to hide away from really making a contribution. He would see them turn up at important team meetings, but the rest of the time he wondered what many of them were doing, and suspected that some were not pulling their weight.

Despite his negative experience with COBE, Wilkinson too was very aware of the limitations of properly probing the size distribution of the CMB fluctuations from anywhere but space. COBE had not been able to make measurements at angular scales less than 7°, and the work being done in the early 1990s from the ground and balloons was starting to probe the fluctuations at angular scales of 0.5° and less, but this still left a massive gap which Wilkinson realised could only be done properly from space. So, when he bumped into Page in the corridor that day, Wilkinson asked Page whether he would be interested in getting involved in a CMB satellite, and Page immediately responded that he would. Both of them sat down and sketched together some ideas for a new CMB satellite, and sent their thoughts off to NASA in what they called a 'white paper'. They named the satellite the *'Princeton Isotropy Experiment'* (PIE), but both of them knew that for several reasons it didn't stand much chance of success.

One of these is that NASA had not even made a call for proposals, which is its first step in eliciting interest in future satellite missions from the astronomical community. In addition, NASA was facing a significant cut in its funding from the US Congress after a few well publicised screw ups. One of the most prestigious NASA missions, the Hubble Space Telescope, had been launched in 1990 at a cost of some $2 billion, only for astronomers to discover that the primary mirror had not been ground correctly and the most costly telescope in history was producing fuzzy images. Following on from this huge embarrassment, NASA's reputation was further diminished by the loss of the Mars Observer probe. Launched in September 1992, at a cost of some $1 billion, it was scheduled to perform a manoeuvre to insert itself into orbit around Mars on the 24th of August 1993. However, on the 21st of August all communication with the satellite was lost. Congress was losing faith in NASA, and its response was to slash the space agency's budget.

Knowing that their 'white paper' stood little chance of meeting with any interest from NASA headquarters, Wilkinson would often refer to the project as *'PIE in the sky'*. He and Page realised that to stand a more realistic chance of success, they would have to work with NASA in a collaboration. After Wilkinson's unpleasant experiences of working with Goddard, they decided instead to approach one of the other NASA facilities, and the most obvious one to choose was the Jet Propulsion Laboratory (JPL) in California. NASA's JPL is a little different from its other centres, although financed by NASA it is actually operated and run by Caltech, its entire purpose is to design, build and operate space missions. Unlike Goddard, whose space missions were mainly in orbit about the Earth (the Hubble Space Telescope, COBE, and many other less well known satellites), JPL had launched many satellites to visit other parts of the Solar System, including highly successful missions such as the Mars Viking landers and the Pioneer and Voyager space probes which had been the first to visit the outer parts of our Solar System.

Wilkinson approached a small group of people at JPL who had also been involved in COBE. They started talking about a Princeton-JPL collaboration, and even gave the proposed satellite a name, the *'Primordial Structure Investigation'*. Wilkinson laid out his requirements to the JPL people, chastened by his experiences of working on COBE. He would collaborate with JPL as long as it was on his own terms, the team had to be small with each team member given plenty of responsibility, and with the scientists involved at every stage of the project. Wilkinson did not want to be involved at the planning and design stage and then see the project handed over to engineers to do the fabrication and testing of the hardware. In fact, Wilkinson stated that he wanted the instrument itself to be built at Princeton, with JPL being responsible for the satellite part of the mission. He wanted it to be a true collaboration, not a JPL project. But, the project did not get much further than giving a name to the potential satellite; negotiations broke down and Wilkinson and Page found themselves almost back at step one again.

Around this time, in the summer of 1993, Chuck Bennett at Goddard had heard that Wilkinson was interested in doing another CMB satellite project. Bennett had been the deputy on the COBE DMR project, working under Smoot. It had been a baptism of fire for Bennett, as when he was appointed deputy of the DMR experiment in late 1986 he was fresh from finishing his Ph.D. at MIT. With Smoot based at Berkeley, Bennett was the most senior person in the DMR team working at Goddard, and so had to take on many of the duties normally associated with the leader of a project rather than its deputy. Sometimes, when Smoot would visit Goddard from Berkeley to check on the project's progress, he would blow up at some of the engineers whom he felt were not doing things as Smoot wanted them to be done. After Smoot would leave, Bennett would work hard on trying to smooth things over with any engineers who had been given such treatment, as he knew the best engineers at Goddard could easily ask to be reassigned to a different project if they started feeling dissatisfied. Bennett became a skilled manager of the DMR team, and Wilkinson and he got on very well, despite Wilkinson's frustrations with aspects of his COBE experience.

In September 1993 Bennett decided to give Wilkinson a call, and asked him if it was true that he was thinking about another CMB satellite mission. Wilkinson responded that it was indeed true, and so they started talking about some of the details. One of Bennett's main concerns was that he wasn't entirely convinced such an experiment could not be done from the ground. If they were to go into space, what kind of orbit would they want? COBE had been in a fairly low polar orbit about the Earth, but would a future satellite be better off in a different kind of orbit? The proposed orbit would, to a certain extent, dictate which launch vehicle they would need, and this would all factor in to what would be the cost of the whole mission. It was clear to Bennett from the conversation that Wilkinson was not the sort of scientist to have grand plans, Wilkinson wanted to do the simplest thing possible. This was good news given the budget constraints at NASA, its then Administrator

Dan Goldin, who had been appointed in April of 1992, was pushing for the space agency to do things "better, faster, cheaper".

After this initial conversation, Bennett set up a meeting at Goddard to discuss the idea further. Wilkinson, Page and Norm Jarosik attended from Princeton, and from Goddard there was Bennett, John Mather (head of the COBE FIRAS) experiment, and Mike Hauser, who had been head of the DIRBE experiment on COBE and overall COBE Principal Investigator. Finally, also from Goddard, was Dave Skillman, a senior engineer at Goddard who was a wealth of knowledge about space missions. Wilkinson had lots of questions about the satellite. How would it be controlled? How cheaply could the mission be done? Which launch vehicle would be available? Skillman was able to answer all of Wilkinson's questions to his satisfaction, and he and his colleagues went back to Princeton fired up with enthusiasm.

Over the next few months, Bennett and Wilkinson continued to chat via email and on the phone. As the project moved forward Bennett realised that they needed to decide who was going to be Principal Investigator (PI) and head up the project. Bennett was heading up the Goddard part of the project, and Wilkinson was heading up the Princeton end, so in many ways they were co-PIs, but it would make the management of the project easier to appoint an overall PI. In a teleconference which also involved Mike Hauser, Hauser suggested that Bennett be the PI so that Wilkinson could concentrate more on the science, and the design, fabrication and testing of the instruments. Wilkinson was more than happy with this, as he was not keen to deal with the administrative duties that being a PI involved, and so the decision was made. Wilkinson was happy, he found he had formed a good understanding with Bennett and found him easy to work with. At one point he commented to Bennett that the two of them had made more progress in a week than he had made in 18 months of discussions with the JPL people.

Bennett knew by now that, in order for the project to progress, they would need some startup money from NASA. He approached some intermediate level administrators at NASA headquarters to make some preliminary enquiries. At this point the project had no science team, no name, no detailed concept beyond trying to improve on the work done by COBE and the work that was currently being done by balloon-borne and ground-based experiments, and they had no design for the telescope. Bennett called the people he knew at NASA headquarters, said he was thinking about a CMB satellite mission, and what did they think? The reply he got was that they liked the idea, but that there was not much money.

Next he approached the Goddard administration and got a more favourable response. He was given some preliminary funding to explore the concept further, and so Bennett and Wilkinson set about recruiting a team of scientists to put together a detailed proposal. Each of them put together a list of who they would like to involve in the mission, and then they discussed their respective lists with each other. When they had both agreed on a name, they would send out an invitation to that person asking them if they would like to be involved. Bennett later said that not a single scientist who received an invitation turned them down.

Wilkinson's Princeton colleagues Page and Jarosik were on the list, as was Goddard software guru Gary Hinshaw. Hinshaw had joined Goddard in 1990 and had helped analyse the COBE data, his main duties being to write and streamline the software that turned the raw measurements from the COBE satellite into the stunning pictures of the microwave sky that had wowed the astronomical world. Hinshaw, they knew, had the computing expertise to build a virtual satellite on the computer and to simulate data based on different cosmological parameters such as different expansion rates, different geometries, different ratios of normal and dark matter etc. They also both agreed on Steve Meyer, then at MIT but soon to move to Chicago, and Ed Wright who was at UCLA. Both Meyer and Wright had been on the COBE DMR team, and Wilkinson and Bennett knew the two of them well and knew that they would be valuable additions to the team. Finally, they both agreed that they wanted John Mather, who had been PI on the COBE FIRAS instrument, to be part of their team. He agreed, but soon had to leave the team as NASA headquarters asked Mather to become the PI on the telescope it was planning to replace Hubble, something it was calling the Next Generation Space Telescope.

Despite the small infusion of money from Goddard which had allowed them to build a team and do some preliminary planning, Bennett and Wilkinson knew that at some point they would need to make a proper proposal to NASA if the mission was going to go ahead. Yet, at this time, NASA was not making any signs that it was going to invite proposals for the next generation of astronomical satellites, it was still in the process of trying to convince Congress to give it more money, and part of that was to make it clear that they would control costs in future much better than they had done in the 1980s when several of their projects grew to costing over a $billion. Those days were over, and NASA knew it, but it had to convince Congress that it knew how to do space science on smaller and tighter budgets. Bennett, Wilkinson and the team they had assembled continued putting flesh on the bones of their idea for a CMB satellite, in the hope that NASA would soon make an announcement calling for proposals.

In March 1994, Bennett caught wind of some hope. The administrators at NASA headquarters were beginning to talk about a new category of missions that they would call 'MIDEX', which stood for 'mid-level explorer'. There was already an 'SMEX' programme, which was for small explorer missions, but Bennett felt that the MIDEX programme was more suited to their needs. According to the rumours that Bennett was hearing, the MIDEX programme was aimed at missions which could be launched within a year or two of approval, and he felt that this was just about right for the small, focused group that he and Wilkinson had pulled together. It was a stark contrast to the bloated COBE mission, which had taken 18 years from its initial concept in 1974 to its final results in 1992. Even before the MIDEX programme was formalised, Bennett decided to talk to the people at NASA headquarters to get a better idea of the programme, and he even made suggestions of his own to try and tailor it more to what he felt would suit the CMB experiment that he, Wilkinson and their team were hoping to propose.

As NASA was putting together its plans for the MIDEX programme, Bennett and his team kept working on more of the details of their experiment. Bennett realised

6.1 Back into Space

that one important detail they still hadn't pinned down was a name for the proposed satellite. He sent out an email to other members of the group asking for a proposed name, and remembers receiving dozens of replies. He remembers making his own suggestion on the 8th of March 1994, the *'Microwave Anisotropy Probe'*. He recalls that this met with unanimous approval from the other members of the team, as it not only described what the satellite was, but its acronym MAP also highlighted that one of its aims would be to make a map of the whole sky at microwave wavelengths. The team were pleased with the name. Bennett's only regret, he would later state, was that the word 'anisotropy' is so unfamiliar to the general public. People outside of the scientific community tend to neither know what it means, nor know how to pronounce it. Bennett recalls having some MAP T-shirts made up for himself and other members of the team, and when he wore his to library 1 day the librarian came over to him and said "I am a real fan of words. What does that word mean?". So, he explained that it meant 'departure from perfect smoothness'.

Although the MAP team now had a name to their project, they still had not worked out any details of what they hoped to accomplish with their proposed satellite, or how they would do it. In June 1994 the team decided that they were still missing a crucial member, they needed to recruit a first rate theoretician. The person they chose to ask to join their team was Dave Spergel, a theoretician who had joined the staff of the physics department at Princeton after making a name for himself working on a theory known as *'cosmic textures'* [1,2]. This competed with inflation to explain the origin of the cosmic structure so evident in the present day Universe. This cosmic textures theory, developed by Spergel and fellow Princeton theoretician Neil Turok, argued that the cosmic structure had not been due to microscopic quantum fluctuations which became macroscopic due to the rapid expansion of inflation in the very early Universe (the cosmic inflation explanation), but were instead due to defects in the fabric of space-time itself, which were created when the Universe made imperfect transitions from one energy state to the next. This is a little like the lines one sees on a frozen lake, where as the water freezes different parts of the ice freeze slightly differently, leading to defects in the otherwise smooth surface. The pattern of anisotropies which COBE found did not fit the predictions of Turok and Spergel's theory, and with great magnanimity Spergel conceded on the day of the announcement in April 1992 that his competing cosmic textures theory was dead.

After the disappointment of seeing his pet theory annihilated, Spergel hastily arranged a meeting at Princeton to discuss the implications of COBE's DMR results, and what the next steps could be in CMB research. Spergel knew his fellow Princeton colleagues Lyman Page and Dave Wilkinson, so he invited them, and they invited some of the key members of the COBE teams who were based elsewhere, along with some other theoreticians. It was during these talks that Spergel heard Dick Bond talk about the cosmological parameters which could be measured from finding and studying the various acoustic peaks in the CMB's power spectrum, and this ignited his interest into moving into this field of study. When he was invited to join the MAP team, Spergel jumped at the chance.

Soon after Spergel joined the project, in late June 1994 Bennett convened a meeting of the MAP team at Goddard in order to thrash out some of the details of their project. They needed to choose some mission objectives, and this would drive their choices of instrumentation, telescope and orbit for the satellite. The team had just heard that the European Space Agency (ESA) had agreed funding for a CMB satellite of its own, with a scheduled launch in 2001. The mission had received 'key mission' status from ESA, which meant that its budget would far exceed anything the MAP team could hope for. The ESA mission's stated objectives were so ambitious and thorough that the MAP team realised that, if successful, it would pretty much measure everything. In order to stand any chance of being funded by NASA, the MAP team had to propose a less ambitious project that could obtain results before ESA's satellite launched.

Once again it was decided that they wanted to surpass what the DMR on COBE had done in looking at the anisotropies in the CMB, and they wanted to do it at angular scales ranging from the 7° scale that COBE had done down to the 0.25° or less angular scales that ground-based and balloon-bore experiments would be able to accomplish. They needed to span the crucial gap between the two, that they felt only a satellite mission could do. The team decided that they wanted to map the entire sky, just as COBE had done, and Gary Hinshaw's computer simulations had shown that in order to be able to properly remove foreground contamination from the Solar System and the Milky Way, it was necessary to observe at a minimum of five different frequencies.

The team also decided that the primary objective in their maps should not be sensitivity, but rather to keep errors to a minimum. Whichever technology they chose for their detectors, they knew that they would be more sensitive than those on COBE's DMR because technology had advanced so much since the 1980s when those detectors were designed and built. They faced a straight choice between radiometers which used bolometers, or which used an electronic amplifier known as a HEMT (high electron mobility transistor). Bolometers, with which Bennett and Meyer were familiar, work by changing their electrical resistance when a microwave signal hits them because the incoming signal causes the bolometer to warm up. They had the advantage of being more sensitive, but they also had a number of disadvantages. In order to work, they have to be cooled to temperatures close to 0 K, which would mean the telescope would need a tank of liquid helium which would add considerably to the size and mass of the satellite. Also, because they are so sensitive to temperature, any tiny change in their temperature leads to a change in the signal that they measure.

To minimise the errors in their measurements, the team all agreed that the detectors would need to make differential measurements rather than absolute ones. This is what the DMR had done, the 'D' stood for 'differential'. Making differential measurements with bolometers was not easy, because the method for making differential measurements involved making a measurement in one direction, then swinging the satellite around to make a measurement in another direction, then to compare the two and to look at the difference. In the process of swinging the satellite around to make the second measurement, it was quite easy for a bolometer's

6.1 Back into Space

response to change, giving a false indication of a difference in the CMB signal when in fact none was present.

HEMTs could operate at room temperature, and were not subject to the same drifts in response as bolometers. Being able to operate at room temperature meant that there was no need for a bulky and heavy helium dewar. Their lower sensitivity compared to bolometers was more than compensated for by the reduction in errors in their measurements due to their better response stability. The only disadvantage the team could see of using HEMTs was that no one in the team had any expertise in building them, which would mean that the work would have to be contracted out to someone outside of the Goddard/Princeton core group, but it was agreed that this was acceptable as the rest of the integration of the detectors would be done in-house at Princeton by Wilkinson and his team.

Another decision they made that day was what kind of orbit they would like for the satellite. COBE's low Earth orbit had been dictated by it being designed to be launched by the Space Shuttle. As it swung around the Earth over the arctic and Antarctic, it also swung from daytime to night-time on each orbit. This led to large changes in the temperature of the satellite, and because it had to point away from the Sun at all times it could only see about twenty percent of the sky during any given orbit. There was no reason to restrict MAP to a low Earth orbit, so the team kicked around ideas for what they would ideally like. One idea they discussed was to put it in a highly elliptical orbit about the Earth, repeatedly sending it out to towards the Moon and back. In such an orbit, most of the time the Earth would be pretty far away and so would be small in the telescope's field of view. Such an orbit would also provide a more stable thermal environment, it certainly looked promising.

But, an even better idea proposed by Hinshaw was to send it to a place in space known as the *second Lagrangian point*, or 'L2' for short. L2 is one of five points where an object will remain in a stable configuration with respect to the Sun and the Earth. The five points are named after French mathematician Joseph-Louis Lagrange, who discovered the L4 and L5 positions in 1772. The other three, L1, L2 and L3 were actually discovered by Leonhard Euler a few years earlier, but unfortunately for him all five points are named after Lagrange. They are illustrated in Fig. 6.1.

A satellite at L2 will move about the Sun at the same rate as the Earth, taking one full year to orbit. It is located about 1.6 million km (about 1 million miles) further from the Sun than is the Earth, in a straight line beyond the line from the Sun to the Earth, as shown in Fig. 6.1. Normally an object in a larger orbit takes longer to complete its journey about the Sun, this is Kepler's third law. But for the L2 location in space, the additional gravitational pull of the Earth on an object leads to it orbiting the Sun faster than it otherwise would, and so it orbits the Sun with the same 365 day period as the Earth. It is an ideal place to put a satellite for several reasons. Firstly, it is a thermally stable location, unlike in Earth orbit, a satellite at L2 will not move in and out of the Earth's shadow, and so the satellite can enjoy a constant temperature. Secondly, the only part of the sky that the satellite needs to avoid is the direction towards the Sun, every other part of the sky is accessible. At 1.6 million km away, the Earth is also just a small object in the sky, about 0.4°,

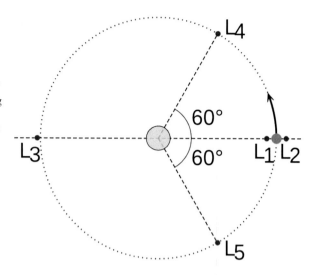

Fig. 6.1 There are five Lagrangian points, the L2 point is where the WMAP satellite would be put. A satellite in an L2 orbit orbits the Sun in the same period of 365 days as the Earth, keeping a straight line between it and the Earth and the Sun. It is an ideal place to located astronomical satellites as it is a cool, stable environment allowing uninterrupted views of most of the sky

similar in size to the Sun. This makes observing the Universe far more efficient than for a satellite in Earth orbit. The only down side is the time it takes to get a satellite to the L2 position, and the larger rocket needed to put it there.

In late July 1994 Bennett heard that NASA were going to issue an NRA (NASA Research Announcement), a way of NASA finding out from the astronomical community what satellite projects they were interested in doing. If they liked the look of it, NASA would give a project a little research money to develop their plans further, before making a decision later on whether a particular project would be chosen to become a mission. Bennett submitted a proposal for the MAP satellite, and was awarded one of the dozen or so feasibility study grants that NASA gave out. However, it then transpired that this NRA was not part of the MIDEX programme that NASA had been talking about, this was being developed along a parallel track. NASA decided that the MIDEX proposal scheme would work in two stages— any team would initially submit a general proposal outlining what they hoped to accomplish with their proposed mission, what the satellite would actually look at and measure and why was it important in an astrophysical sense.

Those proposals which survived this first cut would be given some funding to develop a more detailed proposal for round two. If Bennett was going to commit his time to this two stage selection process, he needed to feel that he had the support of his bosses at Goddard. In August of 1994 he made a presentation to the Goddard administrators, along with others who were also competing for the backing of the Centre's management. His presentation on MAP and its intended goals impressed the Goddard management sufficiently, because Goddard decided to give the project its highest priority.

The first round of proposals for NASA's MIDEX programme were due in the spring of 1995. Eventually NASA would receive some sixty proposals for a whole range of missions from studying the Earth's magnetosphere to looking at the origins

6.1 Back into Space

of galaxies. Bennett got the impression from what he was hearing on the grapevine that NASA would choose at least one proposal to study the cosmic microwave background, as the scientific case was so compelling. But, it was no secret to the MAP team that they were not the only ones proposing a CMB mission. Two other groups were also proposing to do CMB experiments, a team based at JPL and another team based at Caltech. The JPL mission, called 'Primordial Structure Investigation' (PSI) was based on cooled HEMT amplifiers, not unlike MAP's intentions. The Caltech mission, called 'Far Infrared Explorer' (FIRE), proposed to use bolometers. Crucially, in the MAP team's view, neither of these other two missions proposed to make differential measurements.

The first round was dubbed the 'science' round by NASA, and in this round each team had to outline the scientific case for their proposed mission. But, they also needed to give some details of the actual hardware, without having to go into every last detail. They also had to convince NASA that they could deliver their project on time, the launch had to be by April 2001, and within the specified $70 million budget. The level of detail of this first round had to therefore include cost estimates, technical specifications and a management plan including staffing, always the major cost in any science mission. MAP, PSI and FIRE all made it past this first round, testament to the importance NASA placed on CMB experiments. However, Bennett did hear rumours that NASA's least favoured of the three was Caltech's FIRE proposal, which gave the MAP team some hope that they would eventually be the chosen mission.

In writing the second round proposal, the MAP team needed to make sure they understood every potential source of error in the differential measurements that the HEMT microwave receivers would be making. It was important to try and keep the radiometers as cold as possible. Whilst it is true that the HEMT detectors the team had decided to use could operate at any temperature, keeping them cool reduces the thermal noise in the detector system, thus reducing a source of error in the measurements. To keep them cold the detectors would be shielded from the Sun, being in deep space would then ensure they got down to a pretty cold temperature without the need for any refrigerating system. However, in order to scan the sky and to remain steady, the MAP satellite would rotate, meaning the solar shield would be heated differently at different times. The power that the satellite needed to run the detectors and other on-board electronics such as the system that would transfer the data gathered back to Earth all had to be powered, and this power would come from two large solar panels. These panels would also get warm, as would the cabling taking their power to other parts of the satellite. This all had to be modelled, so that the team would know *exactly* how much signal they were expecting to measure from sources that were in fact nothing to do with the CMB.

Another challenge the MAP team faced was in processing their observations. As they were planning to make differential measurements, this would involve comparing each and every observation on the sky with every other observation. The sheer number of comparisons was daunting. With COBE's 7° beam, the number of individual observations in looking at the whole sky was about six thousand, but MAP's beam was much smaller, only two fifths of a degree (0.2°). This would mean that there would be about 3 million individual observations, and comparing each and every one of these to every other one was just not feasible, even with the fastest computers. The team were fully aware of this, and it was Ned Wright who came up with a solution. Whilst driving home one evening from a conference he had attended at the University of California Santa Barbara, Wright came up with an algorithm that he felt would make such a computing task feasible.

As soon as he got in from his drive home, Wright wrote an email outlining his idea to Bennett, Gary Hinshaw and a few of the other team members. The following day, when Hinshaw arrived in work, he saw the email and immediately felt that Wright had hit upon an elegant and clever solution. By this time Hinshaw had produced many simulated maps with different parameters, so he was quickly able to adapt his code to include Wright's algorithm to see how well it worked. Later in the afternoon the team had their answer—Wright's algorithm worked perfectly and dramatically cut the processing time, and would allow the six million individual observations to all be compared to each other in a few hours rather than the previous method's several weeks.

The second round proposal was duly sent off to NASA in December of 1995, then all the MAP team could do was wait to hear whether they had been successful. As the deadline for completing the mission was ticking whilst they waited to hear whether they had been successful or not, the team did not feel they could just sit back and do nothing. They therefore decided to get started on some of the work, including figuring out the detailed design of the HEMT amplifiers, which had to be outsourced to the National Radio Astronomy Observatory (NRAO), based in Green Bank, West Virginia. They also put together over one hundred web pages for the MAP website, including details about the satellite but also many pages of background science to educate the public about what the mission would be attempting to do. They wisely decided it was better to get this done early, so that they wouldn't have to worry about the important public education part of their project when they might have hardware deadlines to meet later on.

Bennett started hearing rumours on the NASA grapevine that one of the three CMB missions would almost certainly be chosen, but he was not able to get a feel whether MAP was the frontrunner or whether FIRE or PSI would be chosen. NASA headquarters would tell them that a decision would be made by a certain date, but then that date would come and go and still no decision had been made. Then, finally, in mid-April 1996, Bennett got a phone call from NASA headquarters to tell him that MAP was formally approved, and had been chosen above PSI and FIRE. The MIDEX programme rules specified that the MAP satellite would need to be launched no later than March 2001, but Orlando Figueroa, head of the Explorer programme office, argued that the team should be told that the launch deadline was August 2000. Bennett argued that this was not achievable, and so eventually a compromise was agreed upon, they would plan to launch MAP on the 7th of November 2000.

Although the HEMT amplifiers were being designed and built at NRAO, they would be integrated with the rest of the detector system in Dave Wilkinson's laboratory at Princeton, in keeping with Wilkinson's desire to have a far more hands-on approach to MAP than he had with COBE. Much of the rest of the detectors and electronics were fabricated at Princeton itself, although some aspects were outsourced to external companies when it became obvious that Princeton's machinists would not be able to deliver all of the components on the extremely tight schedule that the November 2000 launch date dictated. Despite several problems and set backs, by the autumn of 2000 all of MAP's components had been delivered to Goddard and were ready to be assembled with the rest of the spacecraft. For 6 months the satellite was put through a series of brutal tests to ensure that it could withstand the rigours of a rocket launch and being operated in the coldness of space.

Inevitably a few problems arose during this final testing phase. The biggest arose, however, not from a test that was being done at Goddard but from the National Reconnaissance Organization, who build and operate spy satellites. They had experienced failures in a power converter that MAP was also using. There were quite a few of these power converters inside various parts of the satellite's electronics, and so the failures that the reconnaissance office had noticed presented a big problem to the MAP team. The reconnaissance office had decided to delay its own launches of satellites which were using this power converter until they could replace them, but the replacements would not be available right away.

MAP did not have the luxury of waiting, and so a decision was made by Goddard to try and fix the problematic component. The problem had been narrowed down to a component which was inside a metal container, and this component flexed when it was subjected to a vacuum, as would happen in space. When it flexed, it sometimes broke an internal connection, and the power converter failed. Goddard found that it was able to fix the problem if they stiffened the metal container by using epoxy, so each of the power converters was removed from the MAP spacecraft to apply this fix. This enforced fix cost the project three to 4 months, and so the launch slipped past the November 2000 deadline. However, by April 2001, the satellite was fully integrated and tested and was ready to be shipped to Cape Canaveral in Florida for launch.

6.2 The Cosmic Web

In addition to the various ground and balloon-based experiments which had been made in the 1990s, astronomers had also made huge strides in observing the present day structure of the Universe. The ground-breaking observations of Margaret Geller and John Huchra in the late 1970s and 1980s (see Sect. 4.3) had shown for the first time that clusters of galaxies were far from uniformly distributed in space. By the 1990s, advances in technology allowed astronomers to make much larger surveys. The two largest surveys which started in this decade were the 'Two Degree Field Galaxy Redshift Survey' survey (2DFGRS) and the 'Sloan Digital Sky Survey' (SDSS - see Fig. 6.2).

Fig. 6.2 A slice of the sky from the Sloan Digital Sky Survey (image credit NSF/SDSS)

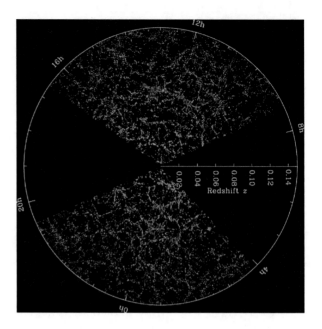

Both surveys took advantage of a new development in measuring the redshifts of galaxies which allowed dozens of individual galaxy redshifts to be obtained simultaneously using fibre optics. After taking an initial image of a particular part of the sky to locate the galaxies' locations, a metal disk which had the same size as the telescope's focal plane would have holes drilled in it at the location of each galaxy, and then a fibre optic from each galaxy would be fed to a spectrograph. This massively increased the speed at which such redshift surveys could be conducted, allowing many more galaxies to be surveyed.

The 2DFGRS survey was conducted using the 3.9 m Anglo Australian Telescope which is based in New South Wales, Australia (a telescope I used in June 1990 to get data for my Ph.D.). The 2DFGRS survey started in 1997 and finished in 2002, during its 5 years of operation the 2DFGRS mapped no fewer than 232,155 galaxies, more than ten times more than the second CfA Redshift Survey, e.g. [3, 4] The SDSS survey was even larger. Using a purpose-built 2.5 m telescope at Apache Point Observatory in New Mexico, about 100 m from the 3.5 m APO telescope that I often use, the SDSS started taking data in 2000 and is still in operation as of 2014. In this time it has mapped the position of over 670,000 galaxies and quasars, mapping some 35 % of the sky, e.g. [5, 6]

Both of these enormous surveys confirm and build upon the findings of the earlier CfA Redshift Survey, that the Universe is highly structured on large scales, with galaxies occupying vast filaments stretching for hundreds of millions of light years, and between these vast filaments are huge voids essentially devoid of galaxies. This structure has come to be known as "the cosmic web".

Armed with these new large scale surveys, theoreticians were able to run simulations of galaxy formation to find which cosmic parameters best suited the observations. The evidence that galaxy formation required the existence of cold dark matter (CDM) had been growing since the late 1970s, but with these larger surveys the theoreticians found that another surprising component was also necessary to obtain the best agreement with the observations.

6.3 A New Standard Candle

The Hubble Space Telescope (HST) had been able to observe Cepheid variables out as far as the Virgo cluster, and in so doing had been able to measure the value of Hubble's constant to a higher level of precision than ever before. In the mid 1990s this 'Hubble key project' reported that Hubble's constant was between 60 and 80 km/s/Mpc, an improvement on the previous range, which was between 50 and 100 km/s/Mpc [7]. Although Cepheid variable stars are intrinsically bright, reaching brightnesses of about 100,000 times the brightness of the Sun, they are still too faint for the HST to see them much beyond the Virgo cluster.

One of the many outstanding questions in astronomy was how much the expansion rate of the Universe had changed with time. Both normal and dark matter would tend to slow the expansion rate down, and so in principle if astronomers could measure the expansion rate in the past they would be able to get an idea of how much matter there is in the Universe and whether the Universe is flat, open or closed. Because gravity is an attractive force, some amount of slowing down is inevitable as the gravitational attraction between clusters of galaxies works against the expansion produced by the big bang. A Universe with less matter in it than the critical density would have a Hubble constant in the past which was very similar to the present day value, but a Universe with more matter in it would have a past Hubble constant which was more than the current value.

Cepheid variables could not be used to measure the distance of galaxies beyond our local Universe, because they were just not bright enough. In the early 1990s, they were the best standard candle we had, but they were not adequate to answer the important question of how much the expansion rate of the Universe had changed in the past compared to now. Astronomers needed to find a standard candle which could be used to much larger distances, but what could they use? Edwin Hubble himself had assumed that galaxies all had the same intrinsic brightness, an assumption which was clearly wrong. Later workers such as Allan Sandage and Gérard de Vaucouleurs had made various other assumptions, including that galaxies of a particular type all had the same intrinsic brightness. Although this assumption was less speculative than Hubble's assumptions, it was unlikely to be correct to the level of accuracy needed to pin down a reliable measurement of the expansion rate of the Universe in the past.

In 1979, before the launch of the Hubble Space Telescope, it was suggested by theoretician Stirling Colgate that Type Ia supernova could be used as a reliable distance indicator [8]. Type Ia supernova are a particular kind of exploding star, produced when a white dwarf accretes matter from a companion until it exceeds its mass limit, a limit known as the *'Chandrasekhar limit'*. This limit, 1.4 times the mass of the Sun, had been worked out by Indian astrophysicist Subrahmanyan Chandrasekhar in the early 1930s as he sailed from his home in India to take up a position at the University of Cambridge with Sir Arthur Eddington. Chandrasekhar had shown that a white dwarf, essentially a huge sphere of carbon atoms, was prevented from further collapse by the electrostatic repulsion between the electrons in the carbon atoms, a force known as *'electron degeneracy pressure'*.

Although electron degeneracy pressure prevents the white dwarf from further collapse, Chandrasekhar worked out that if the mass of a white dwarf exceeds 1.4 solar masses, the force is not strong enough to overcome gravity and the white dwarf will collapse into a neutron star, or possibly a black hole. Most white dwarfs are stable, they gradually cool over time and will fade away from view. But, every so often, a white dwarf will find itself in a binary system with another star. If this star happens to be a red giant star, the white dwarf can pull material off of the swollen red giant. This material will swirl onto the white dwarf, and if this material takes its mass over the Chandrasekhar limit, the white dwarf will explode in what we call a Type Ia supernova.

A Type Ia supernova becomes very very bright, much brighter than a Cepheid variable. As I mentioned above, the brightest Cepheid variables can reach a brightness of some 100,000 times the luminosity of the Sun, the most distant Cepheid observed to date is in the galaxy NGC 3370, a spiral galaxy in the constellation Leo. The distance to the galaxy, as determined from the period-luminosity relationship for Cepheids, is 29 Mpc. This is just a little beyond the distance to the Virgo cluster, which extends from 15 out to 25 Mpc. Type Ia supernovae reach brightness of some 4.5 billion times the luminosity of the Sun, allowing them to be used to distances of more than 1,000 Mpc, about five hundred times further away than the brightest Cepheids!

However, as things stood when Colgate published his paper in 1979, there was too large a variation in the intrinsic brightness of Type Ia supernova to use them as a reliable standard candle; a way had to be found if possible to reduce this scatter based on some observed characteristic of the supernova. A dedicated programme was started to try to observe more Type Ia supernova to study in more detail how they brightened and dimmed, their so-called 'light curves'. Because Type Ia supernovae brighten in a matter of days, in the early 1990s very few had their light curves before maximum brightness measured.

The dedicated supernova search programmes undertaken in the 1990s filled in these gaps, so that by the mid 1990s astronomers had gathered complete light curves for many dozens of Type Ia supernovae [9, 10]. This allowed them to not only determine the range of measured maximum intrinsic brightnesses of the supernovae, but to look for anything in the light curves which might vary between supernovae. What astronomers found was that the rate of decline of the light curve varies between supernovae, and they found that this rate of decline is related to how bright the intrinsic maximum luminosity is. In the 15 days after the peak luminosity, intrinsically brighter supernovae have light curves which decline more slowly, and so by measuring the rate of decline of a Type Ia supernova's light curve from its maximum brightness the variation in intrinsic brightness could be corrected, allowing them to be used as standard candles [11]. This breakthrough now allowed astronomers to use Type Ia supernovae to reliably determine the distance out to distances on the cosmological scale, back to times when the Universe was only about half of its current age, and to therefore determine whether the expansion rate of the Universe was different or not in the past.

6.4 The Most Surprising Astronomical Finding of the Century?

Two teams started using Type Ia supernovae as a reliable standard candle in order to measure the value of Hubble's constant in the past. The two teams were the 'Supernova Cosmology Project' (SCP), based at the Lawrence Berkeley Laboratories and headed by Saul Perlmutter, and the 'High-Z Supernova Search Team' based at Harvard University and headed by Brian Schmidt and Adam Riess. By early 1998 evidence was beginning to emerge that there was something surprising in their findings, later in the same year both teams would announce the remarkable result that they had found the value of Hubble's constant in the past to be less than it is today—the expansion of the Universe is accelerating! [12, 13]

This result sent shock-waves around the astronomical community, it is fair to say that hardly anyone had been expecting such a result. Newspapers around the world picked up on the story, and the journal *Science* made this their *'discovery of 1998'* [14]. What could be causing this acceleration? How could the Universe be expanding faster now than in the past, when gravity should be slowing it down? The simple answer is that we don't know, the effect has been given the name *'dark energy'*, but we don't know what this dark energy is. It is possible that it is the cosmological constant that Einstein had introduced back in 1917 to give his equations of general relativity a static solution for the Universe, or it might be something else.

Although the finding came as a surprise to most of the astronomical community, there were some theoreticians who were less surprised. As the data on the large scale structure of the Universe was coming in from the CfA Redshift Survey, theoreticians were running models to try and reproduce the observed structure on their computers. They found clear evidence of the need for cold dark matter (CDM) in their models, otherwise it was just not possible to produce the observed structure by the present day. Without CDM, galaxies and clusters of galaxies would just form too late to allow time for the structure we see today to form. CDM had, by the late 1980s, become an accepted part of the galaxy formation models, and the observations of galaxy rotation curves and the motions of galaxies within clusters provided additional evidence of the need for dark matter [15–17].

Despite their best efforts, astronomers were not able to find enough dark matter to make the Universe flat, even though theoreticians argued that it was. Combining ordinary matter and dark matter gave an Omega of about 25 %, only a quarter of the value needed for a flat Universe. Where was the missing 75 %? In 1995 Lawrence Krauss of Case Western University in Ohio and Mike Turner of the University of Chicago published an article entitled *"The Cosmological Constant is back"* [18], in which they argued that to produce a flat Universe it was necessary to re-introduce the cosmological constant, usually denoted by the Greek letter Λ, back into our models. In June of 1996, Turner gave a talk at a conference held at Princeton entitled *"The case for Λcdm"* [19], arguing that not only was cold dark matter necessary in any successful cosmological model, but it was also necessary to include the cosmological constant.

It was Turner, Perlmutter and Martin White who coined the term *'dark energy'* in 1998 [20], and so when Perlmutter's SCP team and Schmidt and Riess's High-Z team both discovered that the Universe's expansion was accelerating, this seemed to confirm what a few theoreticians had been arguing for several years. Two lines of observational evidence were now pointing towards a Universe which was flat but with a substantial fraction of dark energy, the observations of the large scale structure of the Universe and the observations of the acceleration of the expansion of the Universe. MAP would be able to measure the power spectrum of the cosmic microwave background to give a third test for the existence or not of dark energy, so its launch in 2001 was eagerly awaited.

6.5 MAP's Detailed Image of the Baby Universe

On the 30th of June 2001, MAP was launched on a Delta II rocket from Cape Canaveral Air Force Base in Florida, just next to the civilian Cape Kennedy launch facility. The satellite used three Earth-Moon loops to help propel it out to its L2 orbit, where it arrived on the 1st of October. During its journey out to L2, the MAP team were able to test that all the spacecraft's systems were working as they expected, so that science observations could begin essentially as soon as it reached its intended location.

In order to scan the sky and to stabilise its orbit and pointing, MAP rotated at just a little bit more than once every 2 min. It orbited around the L2 point in what is called a Lissajous orbit, an elongated figure of eight. This allowed it to always point away from the Sun and the Earth, and to survey the entire sky once every 6 months, so it completed its first full-sky map by April of 2002. Already the MAP team could see from this first scan that their satellite would produce an unprecedentedly detailed image of the fluctuations in the CMB, allowing scientists for the first time to look at the anisotropies over the whole sky from angular scales down to $0.2°$ all with the same instrument. But, one scan of the sky was not sufficient to give the team the level of accuracy they desired, so there would be no official announcement of any results until the instrument had completed its second complete scan.

6.5 MAP's Detailed Image of the Baby Universe

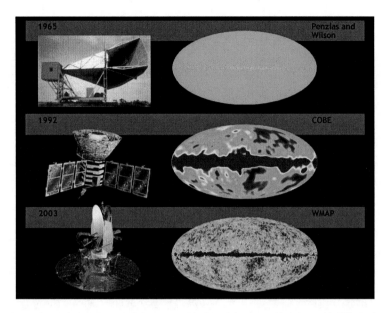

Fig. 6.3 A comparison of the all-sky maps produced by (*top*) Penzias and Wilson in 1965, (*middle*) COBE in 1992 and (*bottom*) WMAP in April 2003 (image credit—NASA)

During this second scan of the sky, on the 5th of September 2002, tragedy struck the MAP team. This was not, however, a tragedy of the astrophysical variety but rather of the human kind. After a 20 year fight against lymphoma, Dave Wilkinson passed away, leaving his colleagues grief stricken at the loss of such a key member of the team, but more importantly a well-loved and well-respected man. The other members of the team decided that they would put in a request to NASA to ask that the spacecraft that he had worked on tirelessly for the previous 6 years be renamed the Wilkinson Microwave Anisotropy Probe (WMAP), and NASA agreed without hesitation.

By February of 2003, the team were ready to release the all-sky observations from one full year of mapping, having surveyed the entire sky twice at five different frequencies [21]. The all-sky image of the anisotropies was stunning in its detail, Fig. 6.3 shows the map in comparison to the COBE map of 1992 and the first ever observations of the CMB by Penzias and Wilson in 1965.

By comparing the relative strengths of the anisotropies at different angular scales, the WMAP satellite was also able to produce the first ever CMB power spectrum which ranged from the large 7° angular scale of the COBE map down to the 0.25° angular scales being probed over limited parts of the sky by the various ground-based and balloon-borne experiments [21]. The first year power spectrum is shown in Fig. 6.4, the error bars are tiny giving an indication of how small the errors in the observations were. As we have outlined before, the position of the first acoustic peak allows us to measure the geometry of the Universe, to determine whether it is

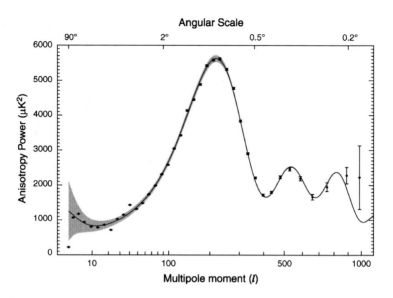

Fig. 6.4 The CMB power spectrum from WMAP's first year of observations. The tiny error bars for all but the smallest angular scales are testament to the very small errors in the measurements (image credit— NASA)

flat, open or closed, or to put it another way whether the critical density Ω_0 is one, less than one or more than one.

The position and strength of the second peak relative to the strength of the first peak allows us to determine the percentage of the density of the Universe which is due to normal baryonic matter. The ratio of the heights of the first three peaks allows us to calculate the abundance of dark matter in the Universe, and so by combining the results of these measurements we can use the power spectrum to also calculate the abundance of dark energy. From the best fit to the data we can also calculate the age of the Universe, the current value of Hubble's constant, and the age of the Universe when matter and radiation decoupled and the CMB was produced. It is amazing to think that all of these key cosmological parameters can be gleaned from the CMB power spectrum, but they can. WMAP's power spectrum after 1 year of observations is shown in Fig. 6.4.

WMAP also had the capability to measure the strength of polarised light. At some point after decoupling, stars and galaxies formed. When the first generation of stars formed they would have re-ionised much of the neutral hydrogen in the Universe, we see evidence for this in the absorption lines produced by hydrogen clouds between us and very distant quasars. This re-ionisation of the early Universe by the first generation of stars would have produced free electrons, and as we have seen electrons are very efficient at scattering, and thus polarising, radiation. Determining when re-ionisation occurred gives as a handle on when the first generation of stars would have formed. Table 6.1 summarises these key cosmological parameters (with their uncertainties) as determined by WMAP after its first year of observations.

6.5 MAP's Detailed Image of the Baby Universe

Table 6.1 Cosmological parameters as determined by WMAP after 1 year [21] and 9 years [22,23] of observations

Parameter	1-Year	9-Years
Age of the Universe (Gyr)	13.74 ± 0.11	13.74 ± 0.11
Hubble constant (km/s/Mpc)	72 ± 5	70 ± 2.2
Age of the Universe at decoupling (kilo years)	379^{+8}_{-7}	$376.971^{+3.162}_{-3.167}$
Redshift of re-ionisation	17 ± 4	10.9 ± 1.4
Baryonic matter density Ω_b	0.046 ± 0.002	0.0463 ± 0.0024
Dark matter density Ω_{DM}	0.224 ± 0.039	0.233 ± 0.023
Dark energy density Ω_Λ	0.750 ± 0.044	0.721 ± 0.025
Total density of the Universe Ω_0	1.02 ± 0.02	$1.037^{+0.042}_{-0.044}$

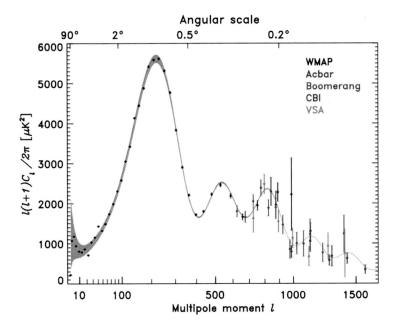

Fig. 6.5 The CMB power spectrum showing 3 years of WMAP observations along with data from Acbar, BOOMERanG, CBI and VSA (image credit—NASA)

In addition to producing an all-sky map at an angular resolution much better than COBE, another important task of WMAP was to bridge the gap in the angular power spectrum between the COBE observations and the various balloon-borne and ground-based experiments which had looked at small patches of the sky on angular scales of typically less than one degree. As Fig. 6.5, the WMAP data are indeed consistent with the data from these other experiments, in important result in and of itself.

Although WMAP was originally funded for 2 years of observations, NASA continued its funding for a further 7 years. This would not have been possible with

Fig. 6.6 WMAP's all sky map of the CMB anisotropies after 9 years of observations [22]. The range of temperatures shown are $\pm 200\,\mu K$ each side of the 2.735 K average temperature (image credit—NASA/WMAP)

either of the other two proposed satellites, FIRE or PSI, because both required active cooling of the detectors which would have been provided by liquid helium. The liquid helium boils away over time, and such cooled detectors typically have lifetimes of three to 4 years before all the liquid helium boils away. The WMAP satellite was finally decommissioned in October 2010, 9 years after its launch. It was moved from its L2 orbit into an orbit where it will orbit the Sun slower than the Earth, taking 15 years to complete fourteen orbits. In this way it will gradually drift away from the Earth.

Each 6 months WMAP re-scanned the entire sky at five separate frequencies, and with each pass it has reduced the errors in the differential measurements, and hence the errors in the derived cosmological parameters. The final data release from WMAP was in December 2012, combining all 9 years of data to produce a final all-sky map of the CMB anisotropies and a final power spectrum [22]. These are shown in Figs. 6.6 and 6.7, but note that the power spectrum is from 7 years of observations as the 9 year power spectrum has not yet been released.

To summarise the findings of WMAP, it found that the Universe is 13.74 billion years old and that decoupling happened when the Universe was 375,000 years old. The current value of Hubble's constant was found to be 70 km/s/Mpc, and the geometry of the Universe was found to be within 0.4 % of being flat. According to WMAP, re-ionisation occurred at a redshift of nearly 11, and finally it confirmed what studies of the large scale structure of the Universe and the evidence for an accelerating Universe had found, that the Universe is mainly dark energy (some 72 %), with dark matter making up some 23 % and normal baryonic matter only 4.6 %. This is summarised in Fig. 6.8.

WMAP has studied the CMB at an unprecedented level of detail, allowing astronomers to obtain precise figures for many cosmological parameters which had been uncertain for many decades. It has added strength to the model that argues that our Universe is currently dominated by dark energy, and that normal baryonic matter

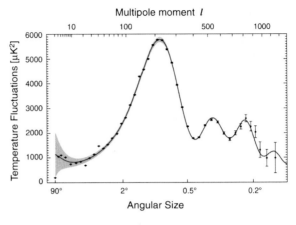

Fig. 6.7 The CMB power spectrum from 7 years of observations [24]. If you compare the error bars to Fig. 6.4 you can see the improvement in the errors with the extra 4 years of observations (image credit—NASA/WMAP)

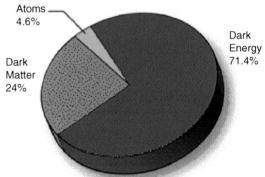

Fig. 6.8 The current make up of the Universe as determined by WMAP. It is dominated by dark energy, with normal baryonic matter making up less than 5 % (image credit—NASA/WMAP)

constitutes less than 5 % of the Universe, with most matter being in the form of cold dark matter. As WMAP was coming towards the end of its 9-year lifetime, the European Space Agency's own satellite Planck would finally go into operation. We discuss its preliminary findings and those of more recent ground-based experiments in Chap. 7, along with a discussion of where cosmological research will go in the future.

References

1. Spergel, D.N., Neil, G., Turok, N.G.: Textures and cosmic structure. Sci. Am. **266**, 52–59 (1992)
2. Gooding, A.K., Park, C., Spergel, D.N., Turok, N., Gott, R. III.: The formation of cosmic structure in a texture-seeded cold dark matter cosmogony. Astrophys. J. **393**, 42–58 (1992)
3. Colless, M.: First results from the 2dF Galaxy Redshift Survey. Philos. Trans. R. Soc. Lond. **357**, 105–116 (1999)

4. Croton, D.J., Colless, M., Gaztañaga, E., Baugh, Carlton M., Norberg, P., Baldry, I.K., Bland-Hawthorn, J., Bridges, T., Cannon, R., Cole, S., Collins, C., Couch, W., Dalton, G., de Propris, R., Driver, S.P., Efstathiou, G., Ellis, R.S., Frenk, C.S., Glazebrook, K., Jackson, C., Lahav, O., Lewis, I., Lumsden, S., Maddox, S., Madgwick, D., Peacock, J.A., Peterson, B.A., Sutherland, W., Taylor, K.: The 2dF Galaxy Redshift Survey: voids and hierarchical scaling models. Mon. Not. R. Astron. Soc. **352**, 828–836 (2004)
5. Sako, M., Bassett, B., Becker, A., Cinabro, D., DeJongh, F., Depoy, D.L., Dilday, B., Doi, M., Frieman, J.A., Garnavich, P.M., Hogan, C.J., Holtzman, J., Jha, S., et al.: The sloan digital sky survey-II supernova survey: search algorithm and follow-up observations. Astron. J. **135**, 348–373 (2008)
6. Anderson, L., Aubourg, É., Bailey, S., Beutler, F., Bhardwaj, V., Blanton, M., Bolton, A.S., Brinkmann, J., Brownstein, J.L., et al.: The clustering of galaxies in the SDSS-III Baryon Oscillation Spectroscopic Survey: baryon acoustic oscillations in the Data Releases 10 and 11 Galaxy samples. Mon. Not. R. Astron. Soc. **441**, 24–62 (2014)
7. Ferrarese, L., et al.: The Hubble constant from the HST key project on the extragalactic distance scale. In: Courteau, S., Strauss, M., Willick, J. (eds.) The ASP Conference Series on Cosmic Flows, Victoria (1999)
8. Colgate, S.A.: Supernovae as a standard candle for cosmology. Astrophys. J. **232**, 404–408 (1979)
9. Hamuy, M., Maza, J., Phillips, M.M., Suntzeff, N.B., Wischnjewsky, M., Smith, R.C., et al.: The 1990 Calan/Tololo supernova search. Astron. J. **106**, 2392–2407 (1993)
10. Hamuy, M., Phillips, M.M., Suntzeff, N.B., Schommer, R.A., Maza, J., Aviles, R.: The absolute luminosities of the Calan/Tololo type IA supernovae. Astron. J. **112**, 2391–2397 (1996)
11. Phillips, M.M.: The absolute magnitudes of type IA supernovae. Astrophys. J. **413**, L105–L108 (1993)
12. Riess, A.G., Filippenko, A.V., Challis, P., Clocchiatti, A., et al.: Observational evidence from supernovae for an accelerating universe and a cosmological constant. Astron. J. **116**, 1009–1038 (1998)
13. Perlmutter, S., Aldering, G., Goldhaber, G., Knop, R.A., Nugent, P., Castro, P.G., Deustua, S., Fabbro, S., Goobar, A., Groom, D.E., Hook, I.M., Kim, A.G., Kim, M.Y., Lee, J.C., Nunes, N.J., Pain, R., Pennypacker, C.R., et al.: The supernova cosmology project: measurements of Ω and Λ from 42 high-redshift supernovae. Astrophys. J. **517**, 565–586 (1999)
14. Glanz, J.: Breakthrough of the year astronomy: cosmic motion revealed. Science **282**, 2156–2157 (1998)
15. Blumenthal, G.R., Faber, S.M., Primack, J.R., Rees, M.J.: Formation of galaxies and large-scale structure with cold dark matter. Nature **311**, 517–525 (1984)
16. Davis, M., Efstathiou, G., Frenk, C.S., White, S.D.M.: The evolution of large-scale structure in a universe dominated by cold dark matter. Astrophys. J. **292**, 371–394 (1985)
17. White, S.D.M., Frenk, C.S., Davis, M., Efstathiou, G.: Clusters, filaments, and voids in a universe dominated by cold dark matter. Astrophys. J. **313**, 505–516 (1987)
18. Krauss, L.M., Turner, M.S.: The cosmological constant is back. Gen. Relativ. Gravit. **27**, 1137–1144 (1995)
19. Turner, M.S.: The case for Λcdm. In: Turok, N. (ed.) Critical Dialogues in Cosmology: Princeton, New Jersey, 24–27 June 1996. Princeton University Press, Princeton (1997)
20. Perlmutter, S., Turner, M.S., White, M.: Constraining dark energy with SNe Ia and large-scale structure. Phys. Rev. Lett. **83**, 670–673 (1999)
21. Bennett, C.L., Halpern, M., Hinshaw, G., Jarosik, N., Kogut, A., Limon, M., Meyer, S.S., Page, L., Spergel, D.N., et al.: First-year Wilkinson Microwave Anisotropy Probe (WMAP) observations: preliminary maps and basic results. Astrophys. J. Suppl. Ser. **148**, 1–27 (2003)
22. Bennett, C.L., Larson, D., Weiland, J.L., Jarosik, N., Hinshaw, G., Odegard, N., Smith, K.M., Hill, R.S., Gold, B., Halpern, M., Komatsu, E., Nolta, M.R., Page, L., Spergel, D.N., Wollack, E., Dunkley, J., Kogut, A., Limon, M., Meyer, S.S., Tucker, G.S., Wright, E. L.: Nine-year Wilkinson Microwave Anisotropy Probe (WMAP) Observations: final maps and results. Astrophys. J. Suppl. Ser. **208**, 20–74 (2013)

23. Hinshaw, G., Larson, D., Komatsu, E., Spergel, D.N., Bennett, C.L., Dunkley, J., Nolta, M.R., Halpern, M., Hill, R.S., Odegard, N., Page, L., Smith, K.M., Weiland, J.L., Gold, B., Jarosik, N., Kogut, A., Limon, M., Meyer, S.S., Tucker, G.S., Wollack, E., Wright, E.L.: Nine-year Wilkinson Microwave Anisotropy Probe (WMAP) observations: cosmological parameter results. Astrophys. J. Suppl. Ser. **208**, 19–44 (2013)
24. Larson, D., Dunkley, J., Hinshaw, G., Komatsu, E., Nolta, M.R., Bennett, C.L., Gold, B., Halpern, M., Hill, R.S., Jarosik, N., Kogut, A., Limon, M., Meyer, S.S., Odegard, N., Page, L., Smith, K.M., Spergel, D.N., Tucker, G.S., Weiland, J.L., Wollack, E., Wright, E.L.: Seven-year Wilkinson Microwave Anisotropy Probe (WMAP) observations: power spectra and WMAP-derived parameters. Astrophys. J. Suppl. Ser. **192**, 6–35 (2011)

Chapter 7
Up to the Present, and Beyond

In 2009 the European Space Agency finally launched the Planck satellite, an ambitious observatory to study the cosmic microwave background in even more detail than had been done by WMAP, including some polarisation measurements. Meanwhile, over the last decade, ground based telescopes at the South Pole and high in the dry Atacama desert of northern Chile have been looking for signs of the 'B-mode polarisation' in the CMB, which it is argued is a direct test of inflation theory. We have moved into an era of 'precision cosmology', where fine details of our theories are being tested, but we are possibly approaching the limit of what we can learn about the Universe from studying the CMB. For future cosmological studies we are developing alternative methods including gravitational wave detectors which will probe the beginnings of time in a different way. Neutrino astronomy is coming of age, and we are increasingly diverting our efforts into understanding the nature of dark matter and dark energy. Some theorists postulate that our Universe is only one of many, that we live in a 'multiverse'. Will we ever be able to know this, and if we can is there a way to learn anything about the properties of other members of the multiverse? Will we ever be able to say anything about the very moment of the beginning of the Universe, or will this always lie beyond our grasp?

7.1 Europe Looks Back to the Dawn of Time

After the success of the COBE mission, the European Space Agency (ESA) started making its own plans to build a European satellite to study the cosmic microwave background. As part of its 'Horizon 2000' scientific programme ESA made a call for proposals and, in 1993, two cosmic microwave background (CMB) satellites were proposed. One was called COBRAS (Cosmic Background Radiation Anisotropy Satellite) which was an Italian/American collaboration. The other one was SAMBA

(Satellite for Measurements of Background Anisotropies), led by mainly French institutions.

A feasibility study in 1994 showed that it would be both more cost effective and lead to a better science mission if the two proposals were merged, and so the COBRAS/SAMBA mission was born. After a detailed assessment phase, the mission was formally chosen in 1996 to be the third medium-sized mission (M3) in ESA's *Horizon 2000* programme. At this point, the mission was renamed Planck, in honour of the German physicist Max Planck who won the Nobel Prize in 1918 for his work in successfully explaining blackbody spectra (see Sect. 3.5), and is so doing discovered the quantisation of radiation. Planck was also the physicist who most championed Einstein's groundbreaking work from 1905, plucking Einstein from the obscurity of the Swiss patent office and leading to Einstein becoming the most famous physicist of the twentieth century.

The main mission objectives of the Planck satellite were

- Perform measurements of Cosmic Microwave Background anisotropies
- Test inflationary models of the early Universe
- Measure the amplitude of structures in the Cosmic Microwave Background
- Determine the Hubble constant
- Perform measurements of Sunyaev–Zel'dovich effect

7.1.1 The Sunyaev–Zel'dovich Effect

I have talked about the first four of these main objectives already, but what is the *'Sunyaev–Zel'dovich'* effect? The Sunyaev–Zel'dovich effect, usually abbreviated to the 'SZ effect,' is named after the two Russian theoreticians Rashid Sunyaev and Yakov Zel'dovich who first proposed it [1, 2]. In 1969 they suggested that the low energy (and thus low frequency) photons from the CMB might interact with the energetic electrons which are associated with the hot gas in clusters of galaxies. This hot gas lies between the galaxies in a cluster and is very hot, at millions of Kelvin, it is one of the main sources of x-ray emission from galaxy clusters.

The electrons in this hot intergalactic gas, which have high energies due to the very high temperatures, may interact with the photons from the CMB through a process known as *'inverse Compton scattering'*. Compton scattering is the high-energy equivalent of the medium-energy Thomson scattering which we discussed in Sect. 5.7; it is when a high energy photon interacts and scatters off of a free electron which has less kinetic energy than the energy of the photon. Normally this involves the higher energy photon giving up some of its energy to the lower energy electron, resulting in the scattered photon having lower energy and thus lower frequency after the scattering process.

Inverse Compton scattering, on the other hand, is the reverse of this. It happens when low energy photons scatter off of high energy electrons. In the process the photons gain energy from the electrons, thus having a higher frequency after they have scattered. Low energy photons from the CMB have to travel through such hot intergalactic gas on their journey to us, and in doing so can experience inverse Compton scattering from the energetic electrons.

Such an interaction will distort the CMB, giving the photons an energy boost and making the CMB appear warmer than it otherwise would. Because this process happens on the scale of clusters of galaxies, it can only have an effect on the CMB's power spectrum at higher angular resolutions. However, it therefore needs to be taken into account when interpreting the observed CMB temperature fluctuations at these smaller angular scales.

Some of the observed fluctuations at small angular scales seen in e.g. Fig. 6.7 for example will be due to the temperature differences imprinted on the CMB at the time of decoupling, 375,000 years or so after the big bang; but some will also be due to the SZ effect which happens after this. In order to distinguish between the two, it is necessary to look at how the CMB fluctuations vary at different angular scales *and* at different frequencies.

Once the two effects have been disentangled, the SZ effect allows astronomers to find clusters of galaxies which may otherwise go undetected. Because the SZ effect is produced by the scattering of photons off of electrons, the resulting photons are polarised (see Sect. 5.7) and the strength of the effect is independent of the cluster's redshift. This is very useful, as it means that clusters at very high redshifts can be detected just as easily as those at low redshifts.

After the 1969 prediction of the SZ effect, it took over a decade for it to be observed for the first time. In 1983 researchers from the Cambridge Radio Astronomy Group and the Owens Valley Radio Observatory (which is part of Caltech) made the first detection of the SZ effect from clusters of galaxies [3]. A decade later, in 1993, the Ryle Telescope, part of the University of Cambridge's Mullard Radio Astronomy Observatory, was the first to image a cluster of galaxies using the SZ effect [4]. Since then, numerous experiments have been made to specifically look for and measure it in more detail. In 2008 the 10-m South Pole Telescope, headed up by John Carlstrom of the University of Chicago and shown in Fig. 7.17, became the first to discover a previously undetected galaxy cluster using the SZ effect [5].

7.1.2 Planck's Specifications

It was decided during the detailed planning phase of Planck that to undertake its main objectives the satellite would be placed in an L2 orbit (see Fig. 6.1) and observe at nine separate frequencies, compared to WMAP's five. This would enable it to better subtract foreground emission from the Solar System ('zodiacal light') and the Milky Way galaxy, a crucial thing in order to examine the CMB in the finest detail. The nine frequencies would be split between two separate instruments, the Low Frequency Instrument (LFI) which would observe at 30, 44 and 70 GHz, and the High Frequency Instrument (HFI) at 100, 143, 217, 353, 545 and 857 GHz. The two instruments would use different technologies for their detectors, the LFI used 22 microwave radiometers cooled to 20 K by liquid helium; the HFI used 52 bolometers cooled to just 0.1 K by a helium-3 dilution refrigerator.

The PI for LFI was Nazzareno (Reno) Mandolesi of the *Istituto di Astrofisica Spaziale e Fisica Cosmica* in Bologna, Italy, who had led the original COBRAS proposal in 1993. The LFI consortium consisted of over 23 institutes, including the Danish Space Research Institute, the *Max–Planck-Institut für Astrophysik* in Garching just north of Munich, and even some non-European partners including NASA's Jet Propulsion Laboratory in the USA.

The PI for HFI was Jean-Loup Puget of the *Institut d'Astrophysique Spatiale* in Orsay, France who had led the original SAMBA proposal, and François Bouchet of the *Institut d'Astrophysique de Paris* was the Deputy-PI. The HFI consortium consisted of over 24 institutes, including my own of Cardiff University, Imperial College London, Cambridge University's Institute of Astronomy, the National University of Ireland and Caltech and Stanford University in the USA. Peter Ade, one of my colleagues in the department here in Cardiff, was UK Instrument Scientist for HFI. Peter has a long and distinguished career in sub-millimetre and millimetre astronomy, and his involvement in Planck came towards the end of his (official) academic career, as Peter is now "retired" (in the sense that academics retire; Peter is still in work every day doing research, but no longer has any involvement in teaching or any other non-research work in the Department).

Peter did his Ph.D. at Queen Mary College, part of the University of London, completing it in 1973. After working as a post-doctoral researcher in the same department, Peter was appointed to a lecturing position at the College, and then later went on to head-up the astronomical instrumentation group there. He became, amongst other things, one of the World's experts in fabricating filters for use in infrared and millimetre astronomy, including developing techniques to finely tune such filters to atmospheric windows when used on ground-based telescopes. I first worked with Peter whilst I was at the University of Chicago, I approached him to make the filters and lenses for the far-infrared camera Chicago has been building for SOFIA, the Stratospheric Observatory For Infrared Astronomy, a project I have been working on since 1997.

In 2001, the decision was made by Peter and other senior members of the group to relocate the entire Queen Mary Astronomical Instrumentation group to Cardiff, thus creating one of the largest and best astronomical instrumentation groups in the UK. This move was made just before my own return to Cardiff from Chicago, and for a couple of years I worked within this group as senior post-doctoral research assistant on a ground-based far-infrared camera that Cardiff was building. In addition to his involvement in Planck, Peter has also been involved in many other experiments, including BICEP2 (see Sect. 7.4.2), and ACTPol.

7.1.3 How ESA Differs from NASA

It is interesting to compare the size of these two groups involved in LFI and HFI to those who built the single WMAP or even the three COBE instruments; it gives an idea of the sheer size of the Planck mission and also the different way in ESA does its projects compared to NASA. The total cost of the Planck mission has been estimated at about 700 million Euros (about one billion), considerably more than the $150 million cost of WMAP, and more akin to the budget of COBE allowing for inflation.

Most major ESA projects will involve institutes from every ESA country, as they are all contributing to its budget and so their national governments, understandably, insist that some of the work comes to their universities and research centres in return. By the very nature of Europe, this means that its major missions are going to have far more institutes involved than are typically involved in most NASA missions.

The LFI and HFI detectors were built to be able to measure both the intensity *and* the polarisation of the photons, with a sensitivity and angular resolution which would exceed that of WMAP. At the lowest frequency of 30 GHz, the angular resolution of the LFI detector was $33'$, just a little over $0.5°$. At the four highest frequencies, the HFI detector had an angular resolution of only $5'$ (or $0.083°$) a factor of over two better than WMAP. The Planck focal plane showing the HFI and LFI detectors is shown in Fig. 7.1. The improvement in the angular resolution

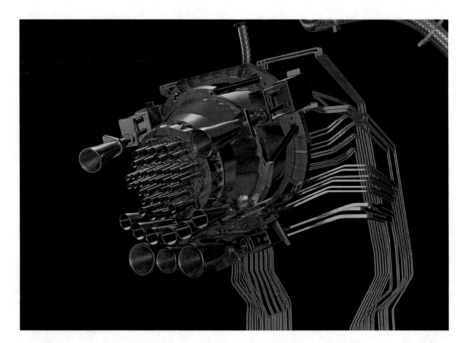

Fig. 7.1 The Planck detectors. The LFI detectors are around the outside, the HFI detectors towards the centre (image courtesy—ESA)

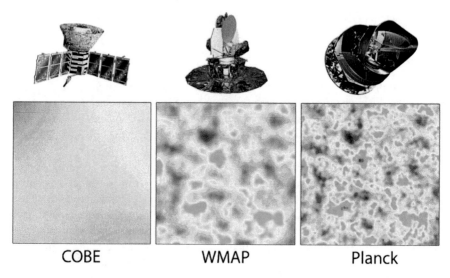

Fig. 7.2 A comparison of COBE, WMAP and Planck resolutions (image courtesy—ESA)

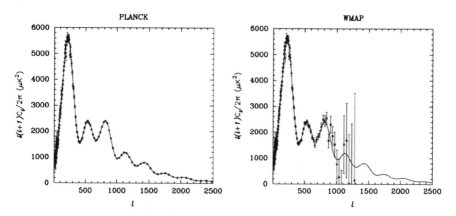

Fig. 7.3 On the *right* is an actual WMAP power spectrum, on the *left* is a simulation of power spectrum that Planck should produce. This can be compared to the actual power spectrum from the first 15 months, which is shown in Fig. 7.7 (image courtesy—ESA)

of WMAP over COBE and Planck over WMAP is nicely shown in Fig. 7.2 which compares the resolution of the anisotropies in the same part of the sky as seen by the three satellites.

Figure 7.3 shows a comparison of the power spectrum of the CMB as Planck expected to measure it (on the left) and as actually measured by WMAP (on the right). This figure, produced in March 2009, shows the large gain in both sensitivity, angular resolution and accuracy that the Planck satellite was expected to achieve.

The contract to launch Planck using an Ariane 5 rocket was signed in December 2005. The decision was made to launch it at the same time a sister ESA satellite,

7.1 Europe Looks Back to the Dawn of Time

Fig. 7.4 The Planck satellite being launched on the 14th of May 2009 aboard an Ariane 5 rocket along with the Herschel Space Telescope (image courtesy—ESA)

an infrared observatory called the 'Herschel Space Observatory'. During launch, Herschel would sit on top of Planck, the two of them enclosed in a structure measuring some eleven metres high, 4.5 m across and with a combined mass of nearly 6 tonnes.

The Planck satellite was originally scheduled to be launched in late 2007, but due to various delays this slipped by some 18 months. It was finally launched along with Herschel on the 14th of May 2009 from the Guiana Space Centre in French Guiana, on the north-east coast of South America (see Fig. 7.4). The two satellites separated soon after launch, and after separation the two have been operated entirely independently of each other. The Planck satellite reached its intended L2 orbit by early July 2009, and the satellite started its initial science observations on the 3rd of July once the spacecraft had confirmed that the HFI instrument had reached its operating temperature of 0.1 K.

Rotating at one revolution per minute, Planck started its first all-sky map on the 13th of August 2009, the initial intention of the mission was for the satellite to complete two all-sky maps in 15 months, but in January 2010 the mission was

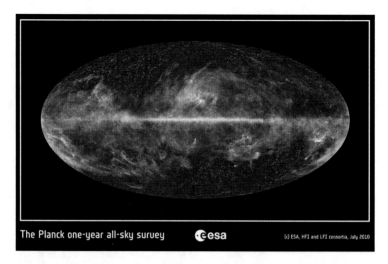

Fig. 7.5 The Planck all-sky map released in July 2010 [6], based on its first year of observations (image courtesy—ESA)

extended by a further 12 months, meaning it would continue observing until at least the end of 2011. In mid-February 2010, it started its second scan of the sky. ESA released the first all-sky map in July of 2010 [6], this is shown in Fig. 7.5. Most of the microwave emission that one can see in this all-sky image is from the Milky Way galaxy, and it shows quite graphically why it is so important to be able to correctly remove this foreground emission to be able to properly study the CMB itself, something that observing at nine different frequencies should allow scientists to do more accurately than ever before.

Planck continued surveying the sky past its 12 month extension and into 2012. In mid-January 2012 the liquid helium cooling the HFI to 0.1 K ran out and so that instrument ceased observations. By this time the satellite had completed five complete all-sky maps, far in excess of the two originally planned. The LFI continued to make observations until early October 2012, whereupon Planck ceased observations and was decommissioned. It was moved away from its L2 orbit into an orbit about the Sun that would leave it drifting away from Earth, orbiting the Sun more slowly than Earth. The satellite was finally switched off on the 23rd of October 2012, nearly three and a half years after its launch.

7.1.4 Planck's 15-Month Map of the Anisotropies and the Power Spectrum

On the 21st of March 2013 ESA released its first all-sky image of the anisotropies in the CMB [7]. The image, shown in Fig. 7.6, is based on the first 15.5 months of

7.1 Europe Looks Back to the Dawn of Time 177

Fig. 7.6 The most detailed map ever of the anisotropies in the CMB, based on Planck's first 15.5 months of data [7] (image courtesy—ESA)

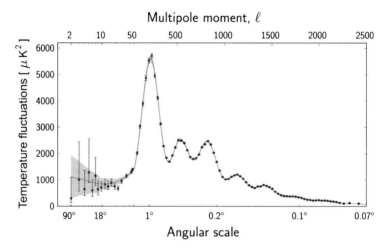

Fig. 7.7 The CMB power spectrum based on Planck's first year of data [8]. The sensitivity and low error of the detectors is such that the second, third, fourth and fifth acoustic peaks can be clearly measured (image courtesy—ESA)

data, and represents the most detailed view yet of the fluctuations in the CMB. The level of detail in this map has allowed cosmologists to determine the cosmological parameters to a level of accuracy never before achievable.

The power spectrum based on this map is also the most detailed yet seen [8]. As can be seen in Fig. 7.7, Planck has measured the strength of the temperature fluctuations at various angular scales ranging from the largest angular scale of 90° down to nearly 0.07°, by far the largest range of angular scales ever studied by one experiment. The sensitivity and small error in the observations are such that the first, second, third, fourth and fifth acoustic peaks in the power spectrum are visible; this is the first time this has been achieved.

Table 7.1 Cosmological parameters as determined by WMAP after 9 years of observations [9, 10] and Planck after 15.5 months [11]

Parameter	WMAP after 9-years	Planck after 15.5 months
Age of the Universe (Gyr)	13.74 ± 0.11	13.813 ± 0.058
Hubble constant (km/s/Mpc)	70 ± 2.2	67.4 ± 1.4
Age of the Universe at decoupling (kilo yrs)	$376.971^{+3.162}_{-3.167}$	370
Redshift of re-ionisation	10.9 ± 1.4	$10.8^{+3.1}_{-2.5}$
Baryonic matter density	0.0463 ± 0.0024	0.04858 ± 0.000726
Dark matter density Ω_{DM}	0.233 ± 0.023	0.26327 ± 0.00682
Dark energy density Ω_Λ	0.721 ± 0.025	0.686 ± 0.020
Total density of the Universe Ω_0	$1.037^{+0.042}_{-0.044}$	0.99785 ± 0.021

The power spectrum from the full mission has not been published yet, and nor have the polarimeter observations of the 'B-mode polarisation' that theoreticians predict will exist in the CMB's light (see Sect. 7.4.1). The Planck consortium hope to release these in late-2014.

Based on the 15.5-month power spectrum, Planck finds the following values for the same cosmological parameters listed in Table 7.1 that WMAP had determined [11].

7.2 Searching for Gravitational Waves

Although we have been able to probe the early Universe beyond the CMB, this has been due to studying the effects the acoustic modes created in the early Universe have on the CMB. But, using EM radiation, there is no way to directly see back beyond the veil of decoupling. Gravitational waves have no such restriction, so in principle we should be able to study the properties of the Universe back to its first moments using them.

7.2.1 What Are Gravitational Waves?

Gravitational waves are ripples in the fabric of space. They were predicted by Einstein as part of his theory of general relativity. In his theory, events which involve extreme gravitational forces (such as two neutron stars orbiting each other, or merging; or the creation of a black hole) will lead to the emission of these gravitational waves. As they spread out from the source at the speed of light, they literally deform space as they pass by, just as ripples deform the surface of a pond as they spread out from a dropped stone. Figure 7.8 shows an artist's impression of gravitational waves being produced as two black holes orbit each other. Figure 7.9 shows a schematic of a laser interferometer used to detect gravitational waves.

7.2 Searching for Gravitational Waves

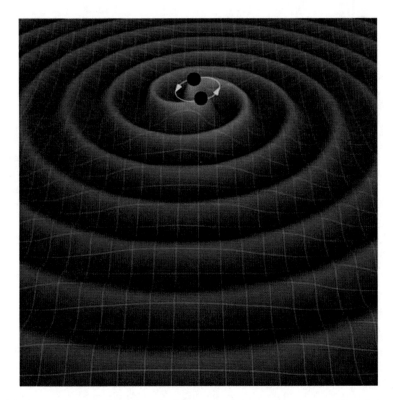

Fig. 7.8 An artist's impression of gravitational waves being produced as two black holes orbit each other

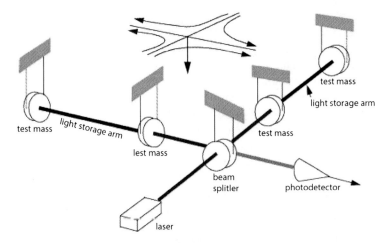

Fig. 7.9 A schematic of a laser interferometer used for detecting gravitational waves

7.2.2 Current Gravitational Wave Observatories

There are several current gravitational wave observatories, all ground-based. These include VIRGO (in Italy) and LIGO (Laser Interferometer Gravitational Wave Observatory) which is in the United States. LIGO is the most sensitive of the current generation of gravitational wave detectors. LIGO actually comprises three separate detectors; one in Livingston, Louisiana and two in Hanford, Washington State. Each of the three separate detectors consists of two long arms at right angles to each other, forming a letter 'L'. The idea behind these detectors is that, if a gravitational wave were to pass the detector, each of the two arms would have its length changed differently by the deformation of space as the gravitational wave passes through. Thus the detectors work on the principle of an interferometer, looking for tiny changes in the relative length of the two arms. And, when I say tiny, I mean tiny. In a 4 km arm they are looking for changes of the order of 10^{-18} m, or about one thousandth the size of a proton!

7.2.3 LISA: Laser Interferometer Space Antenna

ESA announced plans in November 2013 that a space-based gravitational wave observatory will form one of its next two large science missions [12]. The plan is to launch a space-based gravitational wave observatory in 2034, which will have much more sensitivity than any current or even future ground-based gravitational wave observatory. Although NASA also had plans to co-develop a space-based gravitational wave observatory (LISA - see Fig. 7.10), it would seem that, as of 2014, NASA has axed any plans to collaborate with ESA on this project and has no plans of its own in this area.

The ESA plans will be based on the NASA/ESA plans for LISA, which have been on the drawing board for most of the last 10 years. ESA's plan is to build two space-based interferometers, which will be in the form of equilateral triangles as this artist's description shows.

The plan for LISA is to have arms which are 5 million km long! Compare this to the 4 km long arms of LIGO. The changes in the length of a 5 million km long arm would be roughly one million times more than for LIGO, so rather than 10^{-18} m it would be 10^{-12} m, which should be well within the capabilities of the detectors. This means that less energetic events than the collapse of a 10-solar mass star into a black hole would be detectable by LISA. All kinds of astrophysical events which involve large changes in gravitational fields should be detectable by LISA, including the afore-mentioned creation of black holes, but also the merging of neutron stars, and even the merging of less massive stars.

But, possibly most exciting is the opportunity that gravitational waves provide to probe the very earliest moments after the Big Bang. With normal electromagnetic radiation (light, X-rays, infrared light etc.), we can only see as far back as about

Fig. 7.10 An artist's impression of LISA, the "Laser Interferometer Space Antenna". Each interferometer will consist of an equilateral triangle, with each side 5 million km in length (image credit—ESA)

370,000 years after the Big Bang, when the CMB was produced. As we have said before, prior to this time, the Universe was opaque to EM radiation of any wavelength, because it was full of unbound electrons, and the photons just scattered off of them and were unable to get anywhere. Crucially, however, the Universe prior to when the CMB was produced was not opaque to gravitational waves, as they are not electromagnetic and do not scatter off of electrons. Gravitational waves therefore provide a way for us to see back before the CMB, and give us a unique way to learn about the conditions of the Universe in its earliest moments.

7.3 Neutrino Astrophysics

Neutrinos provide another way to observe cosmic phenomena, and provide several advantages over studying objects using EM radiation.

7.3.1 What Are Neutrinos?

Neutrinos are amongst the most mysterious and elusive particles in nature. They were first proposed back in 1930 by Wolgang Pauli to solve a problem to do with radioactive beta decay [13]. In radioactive beta decay, a neutron will turn into a proton, spitting out a high energy beta particle (which is actually an electron) from the nucleus. Experiments showed that the energy of these electrons varied, which seemed to violate the principle of the conservation of energy.

Pauli suggested that the energy was actually being shared between two particles, the electron and a new particle which he dubbed the *neutrino*, which means 'little neutral one' in Italian. However, it was not until 1956 that they were first actually detected [14]. The reason they took so long to detect is that they do not interact very much with matter. They have no electrical charge, so do not feel the electromagnetic force. They have next to no mass (or possibly no mass, that is still an unresolved issue), so do not feel the gravitational force, and they do not feel the strong nuclear force which keeps atomic nuclei together.

The only force they feel is the weak nuclear force. As a consequence of how little neutrinos interact with matter, they can pass through the Earth essentially unimpeded. Every second, billions pass through your body without interacting at all with any of the atoms in your body! However, very rarely, a neutrino will directly strike an atomic nucleus, and this collision enables us to detect them.

Over the last couple of decades increasingly more sophisticated and sensitive neutrino detectors have been built around the World, and we stand on the verge of using neutrinos to make observations that we just could not make using EM radiation.

7.3.2 Neutrinos from the Sun

The Sun gets is power by converting hydrogen into helium in its core, in a process known as the proton–proton chain. During this process, in addition to large amounts of energy being produced, neutrinos are generated (see Fig. 7.11).

The Sun is the strongest source of neutrinos beyond our terrestrial laboratories, but when physicists first started detecting neutrinos from the Sun in the 1960s they discovered a problem [15, 16]. It seemed that the Sun was only producing one third of the neutrinos that calculations predicted, or at least we were only detecting one third. This became known as the solar neutrino problem, and was not solved until the last 15 years.

Neutrinos actually come in three flavours, the electron neutrino, the muon-neutrino and the tau-neutrino. Early neutrino detectors were only sensitive to the electron neutrino, but more recent detectors have been able to detect all three types. They have found that the total number of neutrinos fits in with the number of electron neutrinos being generated in the Sun's core, but about a third of the electron neutrinos are converted to muon neutrinos and a third to tau neutrinos during the 8 min it takes them to come from the core to the Earth.

The theoretical explanation for this, which has not yet been proved, is that neutrinos 'oscillate' between the three flavours. For this to be possible, even the electron neutrino which experiments so far have shown has zero mass would need to have a small mass. The theory of neutrinos and their oscillations is an area of ongoing research, with several unresolved issues.

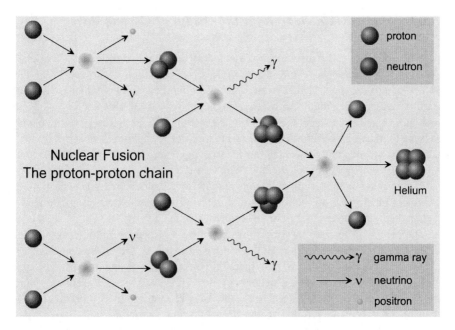

Fig. 7.11 A schematic of the prton–proton chain the in the core of the Sun, which releases a huge number of neutrinos

7.3.3 Supernova 1987A

In February 1987 a star was seen to explode in the nearby Large Magellanic Cloud, a satellite galaxy of the Milky Way. It was seen independently by Ian Shelton and Oscar Duhalde on the same evening whilst both were observing at the Las Campanas Observatory in Chile [17].

This was the first naked-eye supernova since the early seventeenth century, and of course allowed astronomers to study supernovae in detail for the first time. But, 3 h before anyone had seen the supernova, a burst of neutrinos was detected by three separate neutrino detectors, the Kamiokande II detector in Japan, the Irive–Michigan–Brooklyn detector in the USA and the Baksan detector in Russia. These neutrinos (strictly speaking, anti-neutrinos) were produced when the core of the dying star collapsed to form a neutron star. In this process, protons and electrons combine to produce neutrons and anti-neutrinos, in a process known as reverse beta decay. The detection of this burst of anti-neutrinos from supernova 1987A was the first time neutrinos were detected from beyond our Solar System.

7.3.4 The IceCube Search for the Origin of Cosmic Rays

In May of 2012, a large neutrino detector at the South Pole called 'IceCube' (see Fig. 7.12) announced the results of its observations to try to determine the nature of cosmic rays [18]. Cosmic rays are charged particles which strike the Earth's atmosphere, many with energies far in excess of any energy we have managed to achieve even at CERN's Large Hadron Collider. Some 89 % of cosmic rays are high energy photons, about 10 % are high energy helium nuclei (two protons and two neutrons), and the remaining 1 % are high energy electrons. When cosmic rays strike the upper atmosphere they often set off a cascade of reactions in the subatomic particles in the air not unlike the reactions we create in particle accelerators. The electron's cousin the muon was first detected in cosmic rays, long before we had created one ourselves at particle accelerators.

The origin of cosmic rays are a mystery. Although we can quite easily detect these very high energy particles, determining from where they have come is very tricky. The reason for this is that space is permeated with magnetic fields, and charged particles get bent in a magnetic field. So, even if we can accurately note the position in the sky from where cosmic rays have come, it is most likely that their path has been altered during their travel through space, leaving us clueless as to the actual point in the sky that was their origin. This is where neutrinos can help; neutrinos have no electric charge, and so do not get bent by magnetic fields.

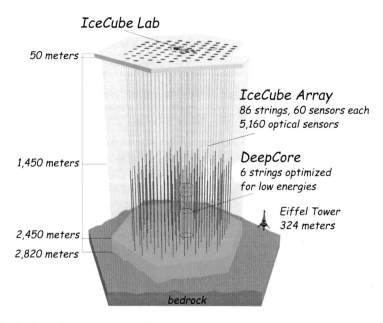

Fig. 7.12 The IceCube neutrino observatory (image credit—NSF)

This of course assumes that neutrinos are also produced in the high-energy events which cause cosmic rays, but this is a reasonable assumption as we see neutrinos being produced in our man-made high energy particle accelerator collisions. Whenever a neutron decays into a proton (beta decay), a neutrino is also produced (strictly speaking it is an anti-neutrino, neutrinos are produced when e.g. a proton and electron combine to produce a neutron in something called 'reverse beta decay').

We can therefore use a 'neutrino observatory' to try and pin-point the positions in the sky of the processes which are producing the cosmic rays. If we could do this we could potentially follow up the neutrino detection with 'normal' telescopes at various wavelengths to see what the nature of the object and event is.

This is what the IceCube neutrino observatory at the South Pole has been trying to do for the last several years. In May 2012 the IceCube team announced that they have detected 28 cosmic neutrinos, which is only the second time that neutrinos have been detected from beyond our Solar system, the first time being when Supernova 1987A exploded. On the down side, as of May 2012 they do not know from which objects these neutrinos had come.

This does, however, bring us a step closer to realising the promise of using neutrinos to better understand the nature of astrophysical objects. In particular, as I described in my previous blog, neutrinos hold the promise of enabling us to understand the origin of high energy cosmic rays. Because the rays themselves are bent by interstellar magnetic fields, tracing their origin is night-on impossible. But, neutrinos are not affected by the magnetic fields, and so should travel to us from their cosmic source in a straight line.

The observatory has also observed a shadowing effect produced by the Moon, which effectively blocks cosmic rays [19]. This leads to fewer cosmic rays showers in the direction of the Moon. As of 2013 the have been three very high energy detections of events in the peta-electron volt range (a peta is 10^{15} or 1,000 trillion which is also called a quadrillion). This is much higher than anything we can produce in the LHC. The IceCube team clearly have a sense of humour, because one pair of high energy neutrinos was dubbed 'Bert' and 'Ernie' after the Sesame Street characters [20]. Later in 2013 an even more energetic neutrino was detected, and so they called it 'Big Bird'! [21].

To date, IceCube is the only neutrino detector in the World which is capable of detecting cosmic neutrinos, but with other neutrino detectors being planned and built, we may indeed soon be entering a new era of neutrino astronomy.

7.4 Can We Learn Much More from Studying the CMB?

Are we reaching the limit of what we can learn from studying the radiation from the CMB? I think the answer is presently 'no', but it may soon be 'yes'. The focus in recent years has turned to measuring the details of the *'B-mode polarisation'* of the CMB.

7.4.1 The B-Mode Polarisation

The main focus of the various CMB experiments in the last half a dozen years or so has been to try and detect and measure what is called the *'B-mode polarisation'* in the CMB. There are two polarisation types in the CMB, the 'E-mode' which is 'curl-free', exhibiting only a gradient (named in analogy to electrostatic field); and the 'B-mode' polarisation which has a 'curl component', (or a 'handedness' to put it another way). Figure 7.13 shows graphically the difference between a vector field which shows only a gradient component and a vector field which has a curl component.

As you can see in Fig. 7.13, the polarisation vectors in the E-mode can be reflected in a line through the centre of the pattern and it would look the same. The vectors in the B-mode, however, cannot. This is what we mean by the vector field of the polarisation vectors having 'curl' or 'handedness'.

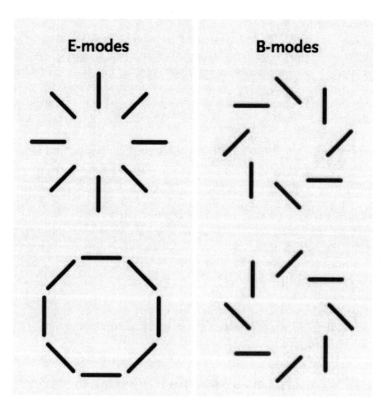

Fig. 7.13 The E-mode polarisation is curl-free, which is to say it doesn't have any 'handedness'. If you were to draw a line through the centre of the E-mode polarisation vectors, they could be reflected in that line and look the same. The B-mode, however, has a curl component, a 'handedness'. If you draw a line through the centre of the B-mode polarisation pattern and reflect the vectors in that line, they do not look the same

7.4.2 The Discovery Made by BICEP2

I mentioned in the Preface the excitement which met the announcement made by the BICEP2 team in March 2014. On the 17th of that month, the BICEP2 team claimed that they had discovered the long-sought after *primordial* 'B-mode' polarisation in the CMB signal [22]. The BICEP2 experiment ('Background Imaging of Cosmic Extragalactic Polarization', the '2' indicates it is the second generation of this experiment) has been making observations of the CMB at the South Pole (see Fig. 7.17) in a small patch of the sky at a high Galactic latitude [23]. The PI on the BICEP2 experiment is Harvard University's John Kovac, who was a Ph.D. student of John Carlstrom at the University of Chicago during the time that I was there. John did his Ph.D. on CARA's DASI telescope, primarily on the observations which led to the first ever detection of the E-mode polarisation in the CMB, as I mentioned in Sect. 5.7.

BICEP2 has been looking for the 'B-mode polarisation' signal in the CMB. Figure 7.14 shows their map of the CMB anisotropies with the E-mode and B-mode

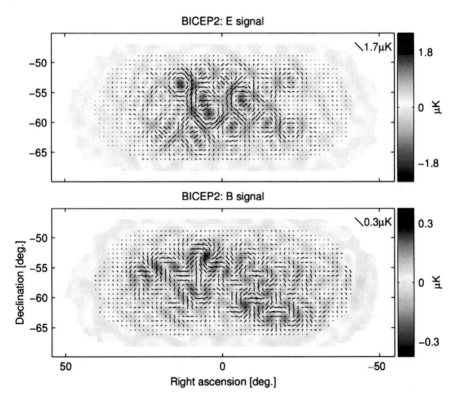

Fig. 7.14 The BICEP2 maps of the E-mode polarisation and the B-mode polarisation in the CMB (image credit—NSF/BICEP2)

polarisation vectors superimposed on the images. The length of each little vector shows the strength of the polarisation, and its direction is the direction of the polarisation. If you look carefully at the polarisation vectors you can see the 'swirl' in the B-mode image. These polarisation features are extremely difficult to detect, as they are at a level of about one part in a billion of the overall blackbody signal.

BICEP2's is the first ever *possible* detection of primordial B-mode polarisation, and hence is the most direct evidence yet that the theory of cosmic inflation is correct. It is argued that this primordial B-mode polarisation comes about due to gravitational waves in the very very early Universe, so the detection of the B-mode polarisation is also claimed to be direct evidence of gravitational waves, which were predicted by Einstein but have never been directly detected before. Figure 7.15 shows the detection made by BICEP2, the solid black dots (with associated error bars) are the signal they detected.

Figure 7.15 requires some explanation. The dashed red line is the signal which would be expected if the observed B-mode polarisation were due to a combination of a mainly primordial component with a small additional component from gravitational lensing of the E-mode polarisation in the CMB by galaxy clusters (which

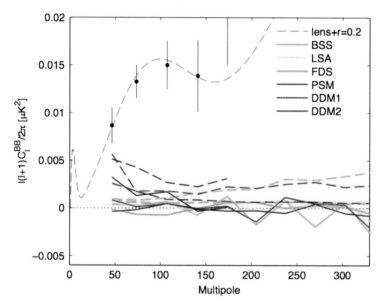

Fig. 7.15 The signal as detected by the BICEP2 experiment is shown by the *solid black dots* (with associated error bars). The *dashed red line* is what would be expected if the B-mode polarisation is dominated by a primordial component, with a smaller contribution from gravitational lensing of the CMB by clusters of galaxies, and as can be seen the signal detected is certainly consistent with this model. The various *dashed curves* at the bottom of the figure show what is predicted to be the contribution due to polarised emission from dust in our Galaxy, as predicted by different models, including two models by the authors (DDM1 and DDM2). If these Galactic dust models are correct, their contribution to the observed signal is clearly negligible (image credit—NSF/BICEP2)

7.4 Can We Learn Much More from Studying the CMB? 189

convert the E-mode to B-mode, this is the B-mode polarisation which was detected by the South Pole Telescope in July 2013). Clearly the BICEP2 measurements fit the dashed red line very well. The other coloured lines towards the bottom of the plot are the contributions to the polarisation signal which would be expected from dust in our Milky Way galaxy according to different models. Of these various models, the DDM1 and DDM2 models are ones which the BICEP2 team have developed, the others are due to other authors. On the face of it, it would seem that the BICEP2 detection is clear and unambiguous, but as I will discuss in Sect. 7.4.4, this is not the case. Since the announcement in March, most of the cosmology community have expressed reservations about the result, and whether the BICEP2 team have done their analysis correctly.

7.4.3 How Do Gravitational Waves Produce the B-Mode Polarisation?

Normally an electron orbits its atomic nucleus in a symmetric orbit, which is illustrated in Fig. 7.16 by 'A'. In reality, the orbit may not be circular as shown in 'A', but it is usually symmetric. But, if a gravitational wave passes through space, then space is alternatively compressed and stretched, as discussed in Sect. 7.2.1 and shown in 'B'. Instead of the electron seeing uniform radiation in all directions, it sees hotter radiation in the direction which is squeezed, and cooler radiation in the direction which is stretched, as shown in 'C'. When these photons Thomson scatter

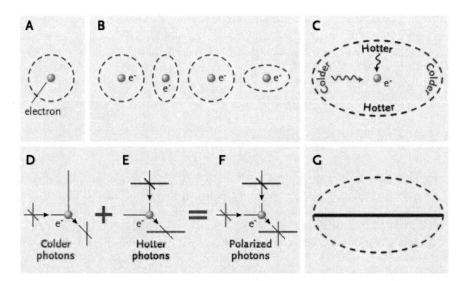

Fig. 7.16 How gravitational waves produce B-mode polarisation

Fig. 7.17 The BICEP2 telescope on top of the Martin A. Pomerantz Observatory building at the South Pole. The 10 m South Pole Telescope is the white telescope to the *left* of the MAPO building (image credit—NSF/BICEP2)

off of electrons, the electrons from the hotter regions will have a stronger electric field—see 'E'—than the ones from the cooler regions ('D'). When they have both scattered off of the electron the scattered EM wave looks like 'F', with the horizontal field stronger than the vertical one. The light will therefore look like it is polarised in the horizontal direction—'G'.

7.4.4 Is the BICEP2 Result Real?

Since the March 2014 announcement by the BICEP2 team, a huge amount of debate has ensued in the cosmology community as to whether their result is real or not. Clearly this is an important question; such a hugely significant discovery should not be accepted by the scientific community without every detail being scrutinised and challenged. This is how science works, new theories or experimental results are not accepted without scrutiny, analysis and questioning by colleagues. As of my writing this in early June, I think it is fair to say that the jury is still out on whether the BICEP2 result is real or not [23]. The obvious caveat is that, so far, this possible primordial B-mode detection has only been made by one experiment, and until it is

7.4 Can We Learn Much More from Studying the CMB?

confirmed by others (including Planck) a good fraction of the community will be reluctant to accept it.

In addition, a number of objections to the validity of the BICEP2 result have been raised, with the main one being that they may not have properly subtracted the polarised signal from dust in our own Galaxy. The BICEP2 experiment made its observations at a single frequency of 150 GHz. This in itself is a weakness, as we know that dust in our Galaxy can be polarised (due to elongated dust grains aligning with magnetic fields in the Galaxy), and so any experiment has to subtract this foreground signal from the observed signal to get at the contribution from the CMB itself.

The subtraction of the Galactic component is difficult if we only observe at one frequency, because disentangling the contributions from foreground dust and from the CMB is really only possible if we observe at several different frequencies. This is because the two come from blackbodies (or, in the case of dust, 'greybodies') at different temperatures. The CMB, as we have seen in Sect. 3.13, is at a temperature of 2.735 K; but Galactic dust is much warmer, by at least a factor of ten. Therefore their blackbody/greybody spectra have different ratios at different frequencies, and this enables us to disentangle them from each other. Without observations at other frequencies, the BICEP2 team had to resort to *models* to do the subtraction.

Unlike Planck and other satellite observations, BICEP2 looked at a small patch of the sky. As I have said before (see Sect. 6.1), ground-based experiments of the CMB are hampered by only being able to observe a small patch of the sky due to variations in the atmospheric emission over large angular scales. BICEP2 wisely chose a patch of the sky which is at a high Galactic latitude to try to minimise the effects of dust from our Milky Way. The patch of sky which they observed is shown in Fig. 7.18. At the time of their writing their paper, there were unfortunately no observations of the emission from polarised dust at 150 GHz, and so the BICEP2 team made a series of assumptions in order to calculate the contribution from the foreground dust, and concluded that its contribution is negligible where they observed.

The various dashed and dotted lines at the bottom of Fig. 7.15 are the contributions from these various models, including two models of their own (DDM1 and DDM2). They argue in their paper that their detected signal is significantly higher than the expected contribution from foreground dust. If their modelling (and those

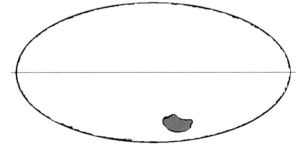

Fig. 7.18 The part of the sky observed by the BICEP2 experiment, as shown in Galactic coordinates with the centre of the Milky Way at the centre of the ellipse (image credit: The Author/ESA)

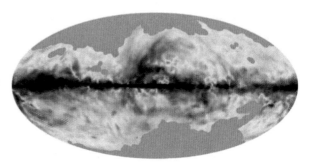

Fig. 7.19 The polarisation map of the Milky Way as released by the Planck Collaboration. Tantalisingly, the part of the sky which covers the BICEP2 region has been masked out. The full analysis will be presented in papers which should be published later this year (image credit: ESA/Planck)

of others) is correct, then indeed the contribution from foreground dust is much smaller than their observed signal.

The modelling, however, *may not* be correct, and this is where the current controversy over the BICEP2 result lies. The Planck collaboration released a map of the emission from polarised dust in our Galaxy in a press release in early May, and the papers describing the details of these measurements have been submitted to peer reviewed astronomical journals. The image ESA released is shown in Fig. 7.19, and in the press release accompanying the image ESA makes direct reference to the BICEP2 claim by saying

> In March 2014, scientists from the BICEP2 collaboration claimed the first detection of such a signal in data collected using a ground-based telescope observing a patch of the sky at a single microwave frequency. Critically, the claim relies on the assumption that foreground polarised emissions are almost negligible in this region.
>
> Later this year, scientists from the Planck collaboration will release data based on Planck's observations of polarised light covering the entire sky at seven different frequencies. The multiple frequency data should allow astronomers to separate with great confidence any possible foreground contamination from the tenuous primordial polarised signal.

It would seem from what has been said by various people that the Planck measurements argue that the contribution from Galactic dust could be as much as 20 %, not the 1–2 % that the BICEP2 team have assumed.

Hopefully the uncertainty will soon be resolved. In addition to results from Planck, there are other ground-based experiments which are looking for the B-mode polarisation, including crucially at other frequencies. The BICEP2 experiment has now been superseded by the Keck Array, which will look for the B-mode polarisation at both 150 and 100 GHz. The South Pole Telescope has started a polarisation experiment (SPTpol) which was the first to report the detection of B-mode polarisation in the CMB, but this was from gravitational lensing converting the E-mode from the CMB to B-mode polarisation, rather than the signature of the primordial B-mode polarisation. It is now, however, hunting for the fainter primordial signal, and is observing at five different frequencies ranging from 95 to 345 GHz. Another experiment, the Atacama Cosmology Telescope Polarisation (ACTPol) will look for the B-mode polarisation at three frequencies, 146, 215 and 280 GHz. In late May they submitted their first paper summarising the first 3 months of observations at 146 GHz of four regions of sky covering a total area of 270 square degrees.

I am confident that the issue of whether the BICEP2 result is true or not will be resolved towards the end of this year (2014), or certainly soon after. In the meantime, at the time of writing, it provides a very good illustration of how science is done. Unlike the judicial system, where a person is assumed innocent until proven guilty, in science a result from a single experiment is assumed to be possibly (or probably?) wrong until (a) other experiments confirm the result and (b) all possible sources of error or a false detection or misinterpretation have been considered. This is the painstaking but necessary process by which science advances, the level of proof required is necessarily very high.

7.4.5 The Future of CMB Research

The progress we have made in our study of the CMB in the last 30 years is truly astounding. It is only just over 20 years ago that COBE discovered the anisotropies in the CMB, and now we are looking for subtle variations in the CMB at the level of one part in a billion! There are quite a few CMB experiments currently in operation with a few about to go into operation in the next year or so. They are all intending to study the B-mode polarisation of the CMB, as measuring it allows us to determine important outstanding questions in cosmology. Some of the ground-based ones have already been mentioned in Sect. 7.4.4, and others are listed below, but I also mention CORE, a proposal to ESA for the next generation CMB experiment from space.

7.4.5.1 Cosmology Large Angular Scale Surveyor (CLASS)

CLASS is a Johns Hopkins University project to locate 4 mm radio dishes in the Atacama desert in northern Chile, at an elevation of 5,200 m. Two of the telescopes will observe at 90 GHz, one at 40 GHz and one at 150 GHz. It will be looking to measure the strength of the B-mode polarisation in the CMB.

7.4.5.2 The E and B EXperiment (EBEX)

EBEX is a balloon borne experiment to study the E and B-modes of the CMB using a 1.5 m telescope and is a collaboration between the University of Minnesota and Brown University, Berkeley and the cosmology group at the University of Paris. The plan was for two flights, with one having been conducted in January 2013 when the balloon flew at an altitude of some 42,000 ft on the polar vortex above Antarctica for some 11 days. The telescope had a 4° field of view and detectors at 150, 250 and 410 GHz. As of writing, there is no mention on the projects web pages of when the second flight will be.

7.4.5.3 QUBIC

QUBIC (A Q&U Bolometric Interferometer for Cosmology) is an experiment to study the B-mode polarisation in the CMB. It is an international collaboration between several universities and research institutes in France, Italy, the UK and the USA. The telescope should being observing in 2016 and will be located at the Franco-Italian 'Dome C' on the Antarctic Plateau.

7.4.5.4 Cosmic ORigins Explorer

The Cosmic ORigins Explorer (CORE) is a proposal to ESA's M4 2015–2025 'Cosmic Vision' programme to build a CMB satellite to observe the B-mode polarisation with a sensitivity of more than one hundred than has been done so far. The overall top-level goals of the mission are

> to probe cosmic origins, neutrino masses and the origin of stars and magnetic fields through a high sensitivity survey of the polarisation of the cosmic microwave background

The satellite would do this by using 15 bands between 45 and 795 GHz. As of the time of writing, this mission has not yet received approval.

7.5 What Is Dark Matter?

As I discussed in Sect. 4.2, the evidence for dark energy has been around since the 1930s, and really gained credence by the 1980s with the discovery of flat rotation curves and the modelling of galaxy formation. I think most astrophysicists today accept its reality. But we are not closer to discovering what dark matter is, although the most likely candidates are weakly interacting massive particles, WIMPs. They are weakly interacting in the sense that the only force they feel is the weak nuclear force, which unfortunately makes them fiendishly difficult to detect. Detection requires a direct collision between a WIMP and an atomic nucleus, just as the detection of the weakly interacting neutrinos is now achieved on a regular basis by detectors such as IceCube.

There are a number of experiments around the World trying to detect dark matter, it is fair to say it is one of the most important current research areas and therefore is receiving a good level of interest and funding. Whether 'axions' or supersymmetric particles like 'neutralinos' exist time will tell, but certainly a good deal of effort is being made to find them.

When students suggest to me that it is a bit crazy to think there are particles like WIMPs which have only been suggested based on theoretical grounds and which have avoided detection for a couple of decades, I remind them that this was the case for neutrinos. They were proposed by Wolfgang Pauli in 1930 but not discovered until 1956!

7.6 What Is Dark Energy?

It is arguably true to say we are closer to figuring out the nature of dark mass than we are to figuring out dark energy. This is the component of the Universe which seems to be causing the expansion of the Universe to accelerate, and it is found to be a necessary component in the models run on computers to reproduce the cosmic web structure seen in our large scale surveys, so called 'Λcdm' models.

The preferred candidate for what dark energy could be is Einstein's cosmological constant, which is now believed to be some kind of negative pressure which exists in space. As space expands, the amount of negative pressure will increase as the amount of space increases,and thus the dark energy is believed to become a larger fraction of the Universe's overall energy budget as time goes on. Several experiments at the moment are trying to determine whether the strength of this negative pressure has changed over the Universe's history, and initial results suggest that it has.

7.7 The Moment of Creation

Will we ever be able to answer the question 'how did the Universe begin?' My own feeling is that no, we will not, and the reason I think this is because even time and space are quantised; that is to say they come in discrete chunks. Because of this, there is a limit to how far back in time we can go.

7.7.1 The Planck Length

The '*Planck length*', usually written ℓ_P, is defined in terms of three fundamental constants, the speed of light c, the 'universal gravitational constant' G (which determines the strength of gravity) and the 'reduced Planck constant' $\hbar = h/2\pi$. It was Max Planck himself who had the idea of combining these constants in this way to produce something with the dimensions of length.

It is written as

$$\ell_P = \sqrt{\frac{\hbar G}{c^3}} \text{ metres} \tag{7.1}$$

If you put in the values of the constants you get that the Planck length is $\ell_P = 1.616 \times 10^{-35}$ m (that is 0.00000000000000000000000000000001616m!). Considering that a proton has a diameter of about 10^{-15} m (a femto metre), the Planck length is a factor of 100,000,000,000,000,000,000 times smaller! To get an idea of just how tiny the Planck length is, we can consider this analogy. The smallest dot that can be seen with the naked eye has a diameter of about 0.1 mm. If we were to scale this dot up to the size of the observable Universe, then the Planck length would scale up to be about the size of the 0.1 mm dot! So, the ratio between the Planck length and a 0.1 mm dot is about the same as the ratio between the 0.1 mm dot and the size of the (observable) Universe!

It is believed that space is quantised at the level of the Planck length, that is to say that it is not possible to measure a distance less than it. The very idea of a distance smaller than the Planck length is, in fact, meaningless, if quantum mechanics is correct. With our current understanding of physics, it would not be possible, no matter how good your 'microscope', to be able to distinguish between something with zero size and something the size of the Planck length.

7.7.2 Quantum Gravity

It is believed by some theoreticians that it is at scales of the Planck length that quantum gravity effects become important, but at present we do not have a theory of quantum gravity. Developing a quantum gravity theory is one of the great theoretical challenges facing physics, as we know that to properly understand gravity we need to unite it with quantum mechanics. At some level, Einstein's general theory of relativity is just an approximation. The leading candidates being used to develop a theory of quantum gravity are 'string theory' and 'loop quantum gravity', but neither has yet been fully developed. 'M-theory' (short for 'membrane theory') is a higher dimensional extension of string theory, and may present an explanation for why there may be more than one 'universe'.

7.7.3 The Planck Time

The *'Planck time'*, denoted by t_P, is related to the Planck length in that it is the time a photon would take to travel the Planck length. It is defined as

$$t_P = \sqrt{\frac{\hbar G}{c^5}} \quad \text{seconds} \tag{7.2}$$

If you plug the values in you get that the Planck time is $t_P = 5.391 \times 10^{-44}$ of a second (that's a decimal point followed by 44 zeros before you get to the 5391!). Again, assuming quantum mechanics is correct, the quantisation of time means that it would be *impossible* to measure a time smaller than this period, in fact a time period shorter than this is meaningless.

This therefore means that we cannot actually say anything about the moment the Universe began, when the time was zero. The furthest back we can go is this Planck time. Although this is an astonishingly small fraction of a second after the beginning, it is *not* the beginning, which is maybe disappointing. My own personal belief is that the beginning, when the time was zero, will always like beyond our grasp. But, on the plus side, we could not distinguish between the beginning and the Planck time in any physical way, they would 'look' the same.

7.8 The Cosmic Onion

In 1894 the physicist Albert Michelson, who in 1887 had helped show that the aether didn't exist with the his colleague Edward Morley, stated [24]

> It seems probable that most of the grand underlying principles have now been firmly established and that further advances are to be sought chiefly in the rigorous application of these principles to all the phenomena which come under our notice. An eminent physicist has remarked that the future truths of physical science are to be looked for in the sixth place of decimals.

It is supremely ironic to think that within 6 years of his making this statement Max Planck would find the first piece of evidence for what became quantum mechanics, and a few years after that Albert Einstein overthrew our very concepts of space and time with his theory of Special Relativity. When I came across this statement as a student in the 1980s it amazed me that someone could be so confident ('arrogant?') about the security of our knowledge. But, later in the 1980s Stephen Hawking was to make a similar statement about where we stood towards the end of the twentieth century. I am not sure if Hawking still thinks that we have discovered most of what there is to discover, but if he does I strongly disagree.

7.8.1 The Multiverse

My view is much more in line with Richard Feynman, and that is that the Universe is most probably like an onion in that it has layer upon layer of complexity. As we understand one layer we discover another layer underneath of which we were ignorant before, and it is my suspicion that this will never end, there will always be some new confusion which emerges as the fog clears on the problem we have just cleared up. Some theoreticians think our Universe is just one of many, the so-called 'multi-verse', and if this is the case how will we ever learn about what exists in other universes?

I may be wrong, but I don't think there will ever come a time when we can say that we understand everything there is to understand about Nature, the Universe, or physics. But, I am sure there are some who would disagree with me. But what I do know with virtual certainty is that I won't be around long enough to know whether I'm right or wrong!

Concluding Remarks

It is quite astounding to sit back and think of the progress we have made in our understanding of the Universe in the last 100 years. We have gone from Kapteyn's Universe to detecting the tiniest twisting of a light which originated when the Universe was only about 370,000 years old, a twisting which probes back to a fraction of a second after the beginning of the Universe. It is impossible to imagine that our progress over the next century will be as dramatic, but who is to say whether we will know by the end of this Century the nature of dark matter, dark energy and whether we live in a multiverse or not?

References

1. Sunyaev, R.A., Zel'dovich, Ya.B.: Distortions of the background radiation spectrum. Nature **223**, 721–722 (1969)
2. Sunyaev, R.A., Zel'dovich, Ya.B.: The observations of Relic radiation as a test of the nature of X-ray radiation from the clusters of galaxies. Comments Astrophys. Space Phys. **4**, 173–178 (1972)
3. Birkinshaw, M., Gull, S.F., Hardebeck, H.: The Sunyaev–Zeldovich effect towards three clusters of galaxies. Nature **309**, 34–35 (1984)
4. Jones, M., Saunders, R., Alexander, P., Birkinshaw, M., Dilon, N., Grainge, K., Hancock, S., Lasenby, A., Lefebvre, D., Pooley, G.: An image of the Sunyaev–Zel'dovich effect. Nature **365**, 320–323 (1993)
5. Staniszewski, Z., Ade, P.A.R., Aird, K.A., Benson, B.A., Bleem, L.E., Carlstrom, J.E., et al.: Galaxy clusters discovered with a Sunyaev–Zel'dovich effect survey. Astrophys. J. **701**, 32–41 (2009)
6. ESA release of Planck's all-sky map from the 1st year of observations. http://www.esa.int/spaceinimages/Images/2010/07/The_microwave_sky_as_seen_by_Planck. Retrieved March 2014
7. ESA release of Planck's image of the anisotropies in the CMB, based on the first 15.5 months of data. http://www.esa.int/spaceinimages/Images/2013/03/Planck_CMB. Retrieved March 2014
8. The power spectrum based on Planck's first 15.5 months of data. http://www.esa.int/spaceinimages/Images/2013/03/Planck_Power_Spectrum. Retrieved March 2014
9. Bennett, C.L., Larson, D., Weiland, J.L., Jarosik, N., Hinshaw, G., Odegard, N., Smith, K.M., Hill, R.S., Gold, B., Halpern, M., Komatsu, E., Nolta, M.R., Page, L., Spergel, D.N., Wollack, E., Dunkley, J., Kogut, A., Limon, M., Meyer, S.S., Tucker, G.S., Wright, E.L.: Nine-year Wilkinson Microwave Anisotropy Probe (WMAP) observations: final maps and results. Astrophys. J. Suppl. Ser. **208**, 20–74 (2013)
10. Hinshaw, G., Larson, D., Komatsu, E., Spergel, D.N., Bennett, C.L., Dunkley, J., Nolta, M.R., Halpern, M., Hill, R.S., Odegard, N., Page, L., Smith, K.M., Weiland, J.L., Gold, B., Jarosik, N., Kogut, A., Limon, M., Meyer, S.S., Tucker, G.S., Wollack, E., Wright, E.L.: Nine-year Wilkinson Microwave Anisotropy Probe (WMAP) observations: cosmological parameter results. Astrophys. J. Suppl. Ser. **208**, 19–44 (2013)
11. Ade, P.A.R., Aghanim, N., Armitage-Caplan, C., Arnaud, M., Ashdown, M., Atrio-Barandela, F., Aumont, J., Baccigalupi, C., Banday, A.J., et al.: Planck 2013 results. XVI. Cosmological parameters. arXiv:1303.5076. http://arxiv.org/abs/1303.5076. Retrieved March 2014

12. ESA announces its plans for LISA, the Laser Interferometer Space Array. http://www.esa.int/Our_Activities/Space_Science/ESA_s_new_vision_to_study_the_invisible_Universe. Retrieved February 2014
13. Brown, L.M.: The idea of the neutrino. Phys. Today **31**, 23–28 (1978)
14. Cowan, C.L., Jr., Reines, F., Harrison, F.B., Kruse, H.W. and McGuire, A.D.: Detection of the free neutrino: a confirmation. Science **124**, 103–104 (1954)
15. Davis, R., Jr., Bahcall, J.N.: Review paper: solar neutrinos. Bull. Am. Astronom. Soc. **1**, 339 (1969)
16. Bahcall, J.N.: Neutrinos from the sun. Sci. Am. **221**, 29–37 (1969)
17. IAUC4316: 1987A, N. Cen. 1986. 24 February 1987
18. Abbasiab, R., Abdouw, Y., Ackermannap, M., Adamsp, J.: Cosmic ray composition and energy spectrum from 1–30 PeV using the 40-string configuration of IceTop and IceCube. Astropart. Phys. **42**, 15–32 (2013)
19. Boersma, D.J., Gladstone, L., Karle, A., et al.: Moon shadow observation by IceCube. arXiv:1002.4900. http://arxiv.org/abs/1002.4900. Retrieved February 2014
20. Dvorsky, G.: Neutrinos from another galaxy have been discovered in Antarctica. http://io9.com/neutrinos-from-another-galaxy-have-been-discovered-in-a-482689596. Retrieved January 2014
21. Klein, S.: Big Bird Joins Bert and Ernie. http://antarcticaneutrinos.blogspot.co.uk/2013/11/big-bird-joins-bert-and-ernie.html. Retrieved February 2014
22. First direct evidence of cosmic inflation. http://www.cfa.harvard.edu/news/2014-05. Retrieved March 2014
23. Ade, P.A.R., Aikin, R.W., Barkats, D., Benton, S.J., Bischoff, C.A., Bock, J.J., et al.: Detection of B-mode polarization at degree angular scales by BICEP2. Phys. Rev. Lett. **112**, 1–25 (2014)
24. Michelson, A.A.: Speech at the dedication of Ryerson Physics Lab, University of Chicago (1894)

Glossary

Absorption line A dark line on a spectrum. Stars show absorption lines, and the position of these lines can tell us how quickly the star is moving and whether it is moving towards us or away from us.

Anisotropy This is a deviation from being uniform. The anisotropies in the cosmic microwave background were discovered in 1992. They are due to small temperature fluctuations in the background, which in turn are due to density variations.

Arc second A degree is divided into 60 arc minutes, and an arc minute into 60 arc seconds. There are 3,600 arc seconds in a degree, and 360° in a full circle.

Big bang theory The term was coined by Sir Fred Hoyle and refers to the idea that the Universe began in a hot, dense fireball, an idea which was first proposed by Georges Lemaître in 1931.

Blueshift The change in colour of the light from an object towards being bluer if the source or receiver are moving towards each other. It is the opposite of redshift.

B-mode polarisation A particular kind of polarisation which shows a curl component. It is believed that it can have two origins—gravitational waves in the early Universe or gravitational lensing of E-mode polarisation by galaxy clusters.

Cepheid variable A particular kind of variable star. Henrietta Leavitt showed that there is a relationship between how long a Cepheid variable takes to vary and its *intrinsic* brightness, in the sense that a longer period Cepheid is intrinsically brighter. This allows us to use them to determine distances.

Conjunction When the planet is in the same direction in the sky as the Sun it is at conjunction.

Cosmic Microwave Background The radiation left over from the hot big bang, first proposed by Ralph Alpher and Robert Herman in 1948 and discovered by Arno Penzias and Robert Wilson of Bell Labs in 1964.

Cosmic rays High energy particles (mainly protons) which strike the Earth's upper atmosphere and cause a cascade of reactions similar to those produced artificially in particle accelerators. Their origin is currently still a mystery.

Dark energy Comprising some 75 % of the mass/energy of the Universe, dark energy is causing the expansion of the Universe to accelerate. At present we do not know what dark energy is.

Dark matter Matter which has a gravitational effect but is not made up of normal atoms or matter. It is "dark" because it does not interact with electromagnetic radiation nor does it radiate any electromagnetic radiation. At present we do not know what it is, the favoured theory are particles called WIMPs (Weakly Interacting Massive Particles—the "weak" meaning they only interact through the weak force. Dark matter makes up some 20 % of the mass/energy of the Universe.

Deferent In Ptolemy's model of the Universe, the Sun, Moon, planets and stars each moved along a circle (deferent) about the Earth.

Degree There are 360 degrees in a circle. A degree is subdivided into 60 arc minutes, and an arc minute subdivided into 60 arc seconds.

Doppler effect Named after Austrian physicist Christian Doppler, it is the change in the wavelength (or frequency) of waves when there is motion between the source and the receiver.

Electromagnetism James Clerk Maxwell showed in the mid 1800s that electricity and magnetism were part of the same phenomenon, and that light is one example of electromagnetic radiation along with e.g. radio waves, x-rays and microwaves.

Electron The negatively charged particle which orbits the atomic nucleus. It was the first sub-atomic particle to be discovered, by J.J. Thomson in 1897.

Ellipse An elongated circle, an ellipse has two foci. In fact, a circle is a special case of an ellipse where the two foci coincide.

Emission nebula A fuzzy patch of gas in the sky which glows, an example is the Orion nebula. We now know that this is due to the gas in the nebula glowing (fluorescing), just like a fluorescent light works.

E-mode polarisation Polarisation of the CMB which shows divergence but no curl.

Epicycle In order to get agreement with observations, Ptolemy introduced epicycles, which were small circles on top of the deferent. The planet would move on the epicycle, as the epicycle moved around the deferent.

Galaxy A vast collection of stars, what philosopher Emanuel Kant referred to as an "island universe". Galaxies fall into very roughly three types; spiral galaxies, elliptical galaxies and irregular galaxies.

Globular cluster A swarm of about half a million stars which orbit the centre of our Galaxy. Shapley used the observed distribution of globular clusters in the sky to argue that our Sun is not at the centre of the Milky Way.

Homogeneous This means the same in different places.

Hubble constant This is a measure of the expansion rate of the Universe and is usually expressed in units of km/s/Mpc where Mpc is a mega parsec.

Inferior planet A planet which orbits the Sun inside of the Earth's orbit. The two inferior planets are Mercury and Venus.

Isotropic This is when something is the same in all directions. It is the opposite of "anisotropic", the origin of the word "anisotropies".

Kelvin The scale of temperature used in physics, named after Lord Kelvin. 0 K is known as absolute zero, the lowest temperature possible. It is $-273.15\,°C$ or $-459.67\,°F$.

Light year The distance light travels in a year, equivalent to 9,467,280,000,000,000 m. It is often the most useful unit to use when talking about the distances to the stars and beyond.

Microwave The part of the electromagnetic spectrum between the radio and infrared. The cosmic microwave background emits most strongly in this part of the spectrum.

Nebula A non-point like object in the sky. The nebulae which had been categorised in the 17th and 18th centuries were "reflection nebula", "planetary nebula", "spiral nebula" and "emission nebula".

Neutrino A tiny by-product of radioactive beta decay, billions pass through your body every second. They interact so weakly that they can travel right through the Earth without being affected. They were first proposed by Wolfgang Pauli in 1930 and were discovered in 1956 by Clyde Cowan and Frederick Reines.

Neutron The neutral particle in an atom's nucleus. All elements have neutrons, apart from hydrogen. They were first proposed by Ernest Rutherford in 1920 and discovered by his co-worker James Chadwick in 1932.

Neutron star This is the remnant of a high-mass star, about three times the mass of the Sun or more. Stars which form neutron stars will explode as supernovae. The neutron star itself has a mass of up to about three times the mass of the Sun but is only about 10 km across.

Nova A "new star" which suddenly appears. We now know a nova is due to an explosion on the surface of a star, but not an explosion large enough to blow it apart as happens in a Type Ia supernova.

Opposition When a planet is in the opposite direction in the sky to the Sun we say it is at opposition. When a planet is at opposition, the Sun-Earth-Planet form a straight line.

Parsec The distance a star needs to be at to exhibit a stellar parallax of 1 arc second. It is equivalent to 3.26 light years and is the unit usually used by astronomers.

Polarisation When the electric field vector of electromagnetic waves is only orientated in a particular direction (linear polarisation) or moves in a regular fashion (circular polarisation)

Proton The positively charged particle in an atom's nucleus. Ernest Rutherford discovered it in a series of experiments between 1917 and 1919, and named it in 1920.

Radioactivity Accidentally discovered in 1896 by Henri Becquerel, unstable atomic nuclei can spit out electrons (beta-particles) or a pair of protons and a pair of neutrons (a helium nucleus) or a high energy gamma ray.

Redshift When objects radiating light move away from us the light is reddened due to the Doppler effect.

Reflection nebula When dust and gas sit off to the side of a bright star the dust scatters the starlight. It is preferentially the blue light which is scattered, so reflection nebulae appear blue. The Pleiades is a well known example of a reflection nebula.

Spectrum When we split light up into its wavelength components we can look at its spectrum. A rainbow shows the spectrum of visible light, from the longer wavelength red light to the shorter wavelength blue light.

Spiral nebula These nebulae have a distinct spiral pattern. We now know they are spiral galaxies beyond our Milky Way galaxy.

Standard candle This is the term used by astronomers for objects which we believe have a known intrinsic brightness, and so we can use their observed brightness to determine their distance. The term was first coined by Henrietta Leavitt in the 1910s.

Steady State theory This was a competing cosmological theory whose main proponent was Fred Hoyle. Before the discovery of the cosmic microwave background, more cosmologists believed in the steady state theory than in the big bang theory.

Superior planet A planet which orbits the Sun outside of the Earth's orbit, such as Mars, Jupiter or Saturn.

Supernova This is when a star explodes, and are amongst the most energetic events in the Universe. A supernova can briefly outshine an entire galaxy. It was Fritz Zwicky who invented the term "supernova" in the 1930s.

The Great Debate This is the name now given to a debate in 1920 between Harlow Shapley and Herber Curtis which was originally entitled "The Scale of the Universe"

Transit This is when an inferior planet (Mercury of Venus) passes across the disk of the Sun. Transits of Venus are extremely rare, but were used to find the distance to the Sun in the mid 1700s.

Type Ia supernova This is produced by the explosion of a white dwarf star, and today they are used as standard candles enabling us to determine distances to galaxies more than 1,000 MPc away.

Wavelength Waves have peaks and troughs, and the distance between two successive peaks (or troughs) is the wavelength.

White Dwarf This is the remnant of a low-mass star like the Sun. They are about the size of the Earth, but with a mass similar to the Sun.

X-ray Accidentally discovered by Wilhelm Röntgen in 1895, x-rays are an energetic form of electromagnetic radiation. They lie between UV light and gamma rays on the electromagnetic spectrum.